MANUFACTURING AUTOMATION AT

STANDARDIZATION IN CIM SOFT\

WITHDRAWN

Advances in Design and Manufacturing

Volume 3

Earlier published in series

Mechatronics & Robotics, I; P.A. MacConaill, P. Drews, K.-H. Robrock (Eds.)
Realising CIM's Industrial Potential; C. Kooij, P.A. MacConaill, J. Bastos (Eds.)

Manufacturing Automation at the Crossroads

Standardization in CIM Software

Edited by:

Louis-François Pau

Digital Equipment Europe

and

Jan-Olaf Willums

Nordic Manufacturing Forum

Contributions to the EUREKA Project INTEGRA
Sponsored by the

Nordic Industrial Fund

IOS Press

1993

Amsterdam • Oxford • Washington • Tokyo

ISBN 90 5199 137 1
LIBRARY OF CONGRESS CATALOG CARD NUMBER: 93-78479

Publisher:

IOS Press
Van Diemenstraat 94
1013 CN Amsterdam
Netherlands

Distributor in the USA and Canada:

IOS Press, Inc.
P.O. Box 10558
Burke, VA 22009-0558
USA

Distributor in the UK and Ireland:

IOS Press/Lavis Marketing
73 Lime Walk
Headington
Oxford OX3 7AD
England

Distributor in Japan:

Kaigai Publications, Ltd.
21, Kanda Tsukasa-cho 2-Chome
Chiyoda-Ku, Tokyo 101
Japan

LEGAL NOTICE
The publisher is not responsible for the use which might be made of the following information.

PRINTED IN THE NETHERLANDS

CONTENTS

SECTION III

Scandinavian Manufacturing Automation in Real Applications:

SECTION IV

The INTEGRA Project Recommendations: Guidelines for Introducing Manufacturing Platforms

PREFACE

by Jan–Olaf Willums

Computer–Integrated Manufacturing, or CIM, became a "catchword" in the late 80's in Scandinavia. Chief executives and factory mangers in the Nordic countries became concerned when they saw their Japanese, German and American competitors focusing on radically new manufacturing technologies, and introducing concepts that started to revolutionise manufacturing efficiency. Some of the most international corporations, such as Nokia in Finland, began to similar initiatives, others, like Volvo, tried to develop their "Scandinavian version" of new manufacturing philosophies. But the majority of the Nordic manufacturing companies were hesitant, reluctant, or even sceptical to these new trends.

The same uncertainty about strategic direction appeared also in other areas relevant to the industrial future of the Nordic countries. The past strategies of basic research and the educational focus at universities and technical institutes were questioned, and questions about the relevance of past industrial research were raised.

It was in that situation of doubt and confusion that the Nordic Council of Ministers decided to launch a Nordic Technology Year, and gave a wide and ambitious mandate to the Nordic Industrial Fund. This Fund was established in 1973 by the five Nordic governments as a vehicle to develop the technical and industrial cooperation among the Nordic countries. It identifies and initiates Nordic research and development projects, with a specific focus on encouraging a co–operation among industrial corporations and research institutes. But it has also increasingly become an instrument for the Nordic industry to increase the links with the rest of Europe in industrial R&D.

One task the Nordic Industrial Fund has set itself is to help develop a network among people in companies and research institutes across Scandinavia. Over the years, the Fund has provided an unparalleled help in bringing experts from the four countries to work together and exchange views and experiences – which has probably given substantial more direct economic benefit to the corporate "bottom–line" than most other government initiatives. The Fund's very capable management has also understood that industrial cooperation cannot be based on technology alone, and has underlined the management aspect in many of their initiatives.

The Nordic Industrial Fund launched therefore, as part of the Nordic Technology Year in 1988, a special project for Technology Management, in which the 16 largest technology corporations in the Nordic countries participated. The heads of technology and research of these industrial leaders, and many of their specialists met many times to exchange views and define a common vision of priorities. It resulted in the launching of Industrial Clubs in several areas, and by 1991, three such groups had become active and economically self–sufficient: The Nordic R&D Managers Forum, The Nordic Environmental Managers Forum, and the Nordic Manufacturing Forum.

12 companies joined the Nordic Manufacturing Forum, and arranged 12 joint seminars and plant visits over a three year period. In line with the Fund's objective to expand the networking to the rest of Europe, the Forum arranged one annual trip abroad, and visits to manufacturing sites in Germany and Italy became successful expansions of the Nordic objective. Manufacturing pioneers like Siemens, BMW, Fiat, and Olivetti opened their doors for the Forum, and brought their top manufacturing experts to exchange their views and visions on a very personal and engaged basis.

During these trips, the Nordic Manufacturing Forum realised that the introduction of CIM had met as many obstacles and problems abroad as they had experienced at home. And it was decided to see if the common experience among the Nordic countries could be combined with the efforts in Germany, Italy and France to develop "productivity platforms" or standards for CIM that could simplify the introduction of modern manufacturing methods also in Scandinavian industry.

The Nordic Industrial Fund was positive to this idea, and agreed to co–sponsor a three–year effort of the Nordic Manufacturing Forum to develop guidelines for, and exchange experience about how to introduce data-controlled manufacturing automation. The specific objective of the project, which was named INTEGRA, was to encourage the standardisation of CIM-software. The Fund insisted, however, in line with its European policy, that the project should apply for EUREKA status in order to draw on the expertise of other European companies and research institutes.

Since its launch in 1985, the EUREKA initiative has generated research projects with total funding of 10 billion ECU. Over 3000 organisations have been taking part in one or more of the 500 EUREKA projects so far. Manufacturing automation and advanced use of computer technology have been key themes in EUREKA projects. The possibility of bringing the ideas of the Nordic Manufacturing Forum in touch with a European–wide initiative, was therefore fascinating, especially as close contacts were already established with leading companies at the forefront of CIM developments.

Thanks to the backing of the Nordic Industrial Fund, the INTEGRA project was given EUREKA status in record time, and soon the first extended

INTEGRA meetings could be held, with specialists from Italy, France, and Germany, and interested parties in the UK and Greece. The link with these groups allowed the Nordic project to build on already on–going work. This allowed us to introduce the ideas and suggestions by the members of the Nordic Manufacturing Forum directly into the work of others. As a result, the work of the Nordic group is today reflected in one of the major CIM software developments that will soon reach a commercial status.

Another requirement of the Nordic Manufacturing Forum was to spread the experiences gained in the project with the widest possible user group in Scandinavia. This has been done on two levels. Several members of the INTEGRA project summarised the general findings of their work in the book "Materialadministrasjon for Konkurransekraft", released in Norway by Ad Notam Gyldendal. It has become the standard reference book for universities in several Scandinavian countries, is already in its second printing, and is presently being translated from Norwegian to other Scandinavian languages.

The objective of this book is to give a more in–depth view of manufacturing automation, and highlight the results of the INTEGRA project both for the technical specialist and the general user or developed of manufacturing software. We believe, like the Nordic Industrial Fund, that the managerial and human side of introducing CIM is as important as the technical advance in the field. We have tried in this book to find the right balance between these two views, which have also dominated the work in INTEGRA.

The INTEGRA project would never have succeeded without the vision of the Nordic Industry Fund, and we would therefore like to dedicate the book to this organisation, and especially its managing director Per Gjelsvik and the programme manager, Peter Gøranson. Their efforts of bringing Nordic manufacturing in closer touch with European mainstream developments will in the long run be recognised as a major contribution to industrial progress in the Northern part of a larger Europe.

Oslo March 1993

Jan–Olaf Willums
past Chairman
Nordic Manufacturing Forum

now Executive Director
International Chamber of Commerce/WICE

ACKNOWLEDGEMENTS

The INTEGRA efforts have involved a large number of enthusiastic persons and institutions. Without their efforts, often on a purely voluntary basis, neither the projects, nor the platforms, nor this book would have been possible. We would first of all like to thank the members of the Nordic Industry Forum, which was the initiating body behind INTEGRA:

Members of the Nordic Manufacturing Forum:
>Alfa Laval, Sweden
>Avenir / Noratom, Norway
>Danfoss, Denmark
>DTH, Denmark (assoc. member)
>Kverneland, Norway
>Jordan, Norway
>Nokia, Finland
>Nordic Enterprise, Norway (Secretariat)
>Sandvik Coromant, Sweden
>S.T. Lyngsø, Denmark
>Valmet, Finland
>Volvo, Sweden

For their sponsorship and financial support, we would like to thank the Nordic Industry Fund project, which made it possible to launch the EUREKA project with the following members:

INTEGRA PROJECT MEMBERS UNDER EUREKA:
>AVENIR / Noratom A/S, Oslo, Norway
>DANFOSS A/S, Nordborg, Denmark
>IKOSS GmbH, Stuttgart, Germany
>ITP Automazione SpA, Torino, Italy
>Nordic Manufacturing Forum / Nordic Enterprise A/S, Oslo, Norway
>SEMA METRA SA, Paris, France
>VALMET Oy, Helsinki, Finland

Furthermore, we would like to thank Dr. Einar Fredriksson, Director of IOS Press in Holland, who has made it possible to spread the experiences of the INTEGRA project world-wide through the publishing of this book.

J.O. Willums, L.F. Pau
Editors

INTRODUCTION

Louis-François Pau
Digital Equipment Europe
Sophia Antipolis, France

The underlying vision behind this book, and the EUREKA INTEGRA project, is that not only the major multinational corporations, but also local manufacturers, and Small and Medium Enterprises (SME's) can be equipped with flexible computerised information systems yielding direct quality, cost and flow control benefits. Computer Integrated Manufacturing (CIM) systems are generally designed for big company needs, with as a result rather specific and costly systems. Also the introduction of such CIM systems then becomes a strategic and sometimes highly emotional issue for the workforce.

Rightfully, the above viewpoint was in 1990 further reinforced by a Japanese (MITI/University of Tokyo) initiative aiming at collaborative research on future manufacturing systems, where the underlying concept was again that basically for example any village could potentially acquire and configure from common platforms, the elements of e.g. a food processing plant catering to their very local wishes.

Scandinavian industry had realised this before, in that in 1988 the Nordic Manufacturing Forum was established with member companies across Scandinavia. Most of these companies were small or medium–sized by international standards, and all operated small manufacturing units. The Nordic Manufacturing Forum soon developed a project idea, called INTEGRA, representing a concept and product recommendation for a common portable low–cost modular CIM system, suitable for end–user customisation. Contacts were made with a diversity of parties, and an international working group with members from France, Italy, Germany and the Scandinavian countries was set up.

In 1990 INTEGRA was granted the EUREKA label, and the joint work could start aiming at specifying such a platform and testing the concept at various manufacturing sites of the Nordic Manufacturing Forum member companies.

The working group soon realised that developing a full set of standards would be beyond both the scope and intention of the group, and focused rather on making recommendations on what elements such platforms should contain, and what criteria they should fulfil. Parallel developments and further analysis took place, aiming at achieving part of the initial vision. Some Nordic

Manufacturing Forum members compared the INTEGRA recommendations against other approaches, some surveyed it in their local economic context, and others went ahead tracking advanced R & D in complementary areas.

The different chapters of this volume all represent parts of the INTEGRA vision, and address the main facets of it. The styles are diverse, thus highlighting that different approaches and argumentation can sustain that same vision.

In the first chapter, Dr. Jan Olaf Willums, one of the founders of the Nordic Manufacturing Forum, describes the reasons for launching the INTEGRA project. He describes how the CIM issue is becoming an important strategic issue for corporations in Scandinavia, and emphasises the entrepreneurial dimension of implementing new visions of productivity in Scandinavia, illustrating the role of consensus for technical developments are carried out and distributed in Scandinavia. Markku Kangas, Vice President of Valmet Oy in Finland, explores then in Chapter 2, and how this concept has had important impacts in Finland.

The aim of the INTEGRA project was to help reduce the barriers to introducing modern manufacturing software in Scandinavian industry. The book has therefore been structured into three main sections: First we analyse the trends in computer integrated manufacturing. We then look at some real application experiences in Scandinavian industry. Finally, we summarise the recommendations that emerged from the INTEGRA project.

In Chapter 3, Professor Hans Andersin places INTEGRA concepts along the path of the historical development of manufacturing automation, from low–level "hard" (hardware based) automation, to "soft" automation. He especially introduces the implications of Process modelling and of a Cost Management System (CMS), together with the features thereof in relation to an INTEGRA–type platform.

In Chapter 4, Prof. L. F. Pau presents the future of the INTEGRA vision, by presenting a scenario and a portfolio of techniques for efficient manufacturing systems and standard. It is especially stressed that much work remains to be done around sensors and metrology systems for greater flexibility at lower costs, around data networks, and around the very design methods for CIM software. In other words, an INTEGRA platform depends on other inputs and media for coordination, which must also obey the same vision.

Chapter 5 describes the Esprit project CIM-OSA, which is very complementary to the INTEGRA initiative, and with which the project has extensive exchanges of experiences during the program.

In Chapter 6, Antonio Camurri, Paolo Franchi, Francesca Gandolfo, and Renato Zaccaria, exemplify the INTEGRA vision through a detailed description of a critical part of the "soft automation", namely process scheduling and the modelling of the overall control of a CIM system. More specifically, a full description is given of a technique whereby to synthesise automatically the real–time CIM system scheduler from high level Petri net models which can easily be adapted to new immediate concerns.

The real applications of these theories in Scandinavian industry are shown in the next three chapters. Jan Storve and Stig Ulfsby define in Chapter 7 the future role of manufacturing platforms for Scandinavian industry. Chapter 8 explores the experiences at Danfoss, the major Danish manufacturer. Stig Ulfsby exemplifies in Chapter 9 the pitfalls and opportunities for CIM in medium sized Norwegian enterprises, and analyses the results of a survey as well as some cases.

The recommendations of the INTEGRA projects are summarised in the next 3 chapters. The Italian member of the INTEGRA project outlines in chapter 10 a proposed guideline for selecting manufacturing software. In Chapter 11, Joseph Shamir and Federico Pisani, as the other Italian partners in the INTEGRA project, have formulated a procedure for implementing CIM and other manufacturing software, based on the experiences in Italian corporations.

Jan–Olaf Willums concludes the book with a view towards the Future, and summarised the recommendations of the INTEGRA project with a call for an international standard for manufacturing software, that could be based on the work initiated by the INTEGRA project partners.

SECTION I:

THE INTEGRA PROJECT AND THE ROLE OF CIM IN NORDIC MANUFACTURING INDUSTRY

Chapter 1

THE INTEGRA PROJECT:

FACTORY AUTOMATION AS A STRATEGIC ISSUE

Dr. Jan–Olaf Willums
Nordic Manufacturing Forum
P O Box 301
1324 Lysaker
Norway

Abstract: This paper describes first the reason for establishing the Nordic Manufacturing Forum, and for launching the INTEGRA project. It highlights the requirements for introducing modern manufacturing concepts to Scandinavian manufacturing, and explains how actual operational experiences were brought into the INTEGRA project to formulate a set of guidelines.

1. INTRODUCTION

Information technology has become an important discipline for the manufacturing industry. However, the complexity of modern production have made manufacturing dependent on a rapidly developing computer–based support technology.

The growth of a multitude of data–solutions, and the use of incompatible products on different factory locations have led to so–called "islands of automation". Such islands may be of considerable individual value, but pose integration problems if one wishes to integrate factory functions or achieve company–wide synergies the objective of Computer–Integrated Manufacturing.

The complexity of the modern factory sets stringent requirements to the systems integrator. Developing customised solutions that can satisfy the quality– and reliability requirements of the modern factory is time–consuming and thus expensive. The experience gained in one project is difficult to transfer to another project unless the integration is carried out by the same team of system integrators.

2. THE NEED FOR STANDARDISED SOLUTIONS.

These problems have been recognised by industry, and several efforts to standardise CIM solutions and make the exchange of experience easier have been launched. The Esprit–initiative CIM–OSA was one of the major international joint effort by 19 computer companies, manufacturers, and systems integrators to define a Reference Architecture from which particular architectures can be derived to fulfil specific needs.

The Nordic Manufacturing Forum has as one of its goal to increase the efficiency of manufacturing productivity, and to spread the acceptance of CIM. The Forum was initiated by the Nordic Foundation for Industrial Development, a group set up by Volvo's chairman in 1981 and backed today by 35 corporations in Scandinavia, and includes major manufacturers in Sweden, Norway, Denmark and Finland.

3. THE INTEGRA PROJECT

The Forum's principle project in the CIM field of computer–integrated manufacturing was to promote a standardisation in factory automation software, that conforms to a common set of stringent industry requirements, and promote the acceptance and utilisation of that common software basis among Scandinavian manufacturers, systems integrators and hardware vendors. By promoting a "quasi–standard" across several automation sectors, the Forum believes that more Scandinavian manufacturers will be more willing to introduce CIM concepts – a necessity for both the larger and smaller Scandinavian manufacturers if they want to remain competitive on a European market.

The basic idea behind a common software platform concept is to extend operating systems, networks, and data base facilities to address the unique requirements associated with the automation of manufacturing processes. Such a platform could allows standardisation of many key software aspects in factory automation. A platform used by a larger group of manufacturers and systems integrators will also allow a better accumulation of experience, and the possibility to exchange such experience between companies. The initiative was therefore called INTEGRA PROJECT, to reflect that it aimed at integrating the experience of many automation specialists.

The working group established under the Nordic Manufacturing Forum recognised early that it would be impossible to meet the needs for integrated automation using off–the–shelf products because of the diversity and complexity of manufacturing requirements and technologies. They suggested therefore to propose a model that would allow to evaluate the appropriateness of reusable software modules within an advanced development environment.

As a software platform should preferably contain experience from actual factory automation solutions, it must develop over time in a milieu exposed to factory automation continuously. The INTEGRA project focused therefore on bringing the experience of actual projects together into a set of guidelines and test cases. Bringing this kind of experience to bear in the Scandinavian manufacturing sector is of importance for the future competitiveness of the Nordic manufacturing industry.

4. PRESENT SOFTWARE NEEDS IN MANUFACTURING AUTOMATION.

The concept of a productivity platform is relatively new in the manufacturing business, but gains acceptance in more and more larger corporations. The main argument for choosing a platform strategy is that the programming tools available on general–purpose operating systems do not address the complex problems encountered in developing software for real–time monitoring and control. A platform provides generic application programs, and a development environment to help the end–users to write their own applications.

Greater benefits accrue from a manufacturing automation strategy that not only integrates automated operations with the manufacturing domain but also integrates the manufacturing domain with the engineering and business domains, which are further down the automation path. Managers of automation projects should therefore look not only at the capability of software to meet the immediate needs of projects, but also to the ability of software to support the trend toward integration within the enterprise.

Chapter 2

CIM – A CHALLENGE OF THE FUTURE OF MANUFACTURING IN THE NORDIC COUNTRIES

Markku Kangas
VALMET Corporation
P B. 155
SF 00131 Helsinki
Finland

Abstract: This Chapter takes a top–down management view of what CIM is about, in terms of the "Simplify–Integrate–Automate" constituents. CIM is also highlighted as a strategy and organizational tool, and not as a technology. The real benefits are listed and examined in this light, as users are also advised to take a step–by–step low–tech approach to CIM for higher returns on it.

1. INTRODUCTION AND DEFINITIONS

Much debate has taken place in the last few years about the nature, consequences and characteristics of Computer Integrated Manufacturing (CIM). However, it is impossible to discuss CIM without first defining it, for the simple reason that there are a wide variety of definitions and viewpoints, starting with purely technical interfacing issues and ending with a total systems approach, which covers MIS, CAD, CAM, Factory Automation and Logistics Systems.

CAM–I, a highly respected international organization, has launched the term CIE", Computer Integrated Enterprise", to emphasize the holistic view. I agree with the CAM–I's CIE definition, but prefer to use the word CIM as having the same meaning in this Chapter. I would therefore like to start by defining CIM as follows:

CIM is a broad conceptual term, which covers both strategic and operational issues in the whole supply chain – starting from customer orders, covering manufacturing and purchasing and ending in the distribution management.

To put it in another way:

There is a need to integrate Manufacturing Systems, such as CAM and FMS, with Information Systems, such as MRP II and CAD, and with "People Systems", such as managers, engineers, operators and organizations.

My viewpoint in this paper is a conceptual one, not technical. The conceptual approach is invaluable because it serves to capture, even if imperfectly, the complexity of systems, and thus contributes to better operational decisions and implementations.

2. TODAY'S MANUFACTURING ENVIRONMENT

In today's world of international and globalized competition it is not possible to be concurrently the lowest cost, most flexible, most responsive and most innovative manufacturer. These attributes are contradictory.

However, new methodologies and new technologies may lead to the creation of new factories, which provide a well balanced, continuous flow facility, combining a high degree of flexibility and low cost. It is predicted that CIM and flexible manufacturing technologies already now dramatically change the way companies do business. CIM is heralded as the key to increased productivity and quality so the Western countries have a change to surpass the Japanese. The often heard battle–cry is "Beat them with technology – or move over". Another way to say the same is "Automate, Emigrate or Evaporate".

The Japanese have challenged the traditional practices in manufacturing and have created tightly co–ordinated and synchronized production, with small lot sizes and short lead times. The lot sizes are small, but the total volume is high. This can be called "economy–of–scope" to distinguish it from "economy–of–scale". *Economy–of–scope* is based on the assumption that a certain *range* of products can be manufactured with the same manufacturing system cost–effectively in small lots, in which case it is advantageous to increase the total quantity by increasing the number of product variations.

The experience of implementing advanced manufacturing technologies and CIM suggest that the key issue is not the technology but the intelligent application of the concepts pertaining to the best manufacturing practices, and then the mastery of the process as a part of a much larger organizational system. The focal point is not to enhance the outdated mass–production facilities by providing more automation and more flexible transfer lines. The focal point is to refocus the manufacturing system so that it has mental and physical preparedness to exploit flexibility, shorter runs and shorter product life cycles.

My conclusion is that the future CIM implementations will create much better results when they are based on Japanese–style production philosophies, i.e. JIT and TQC, and when they follow the approach of "Simplify, Integrate, Automate".

2.1. SIMPLIFICATION AND ELIMINATION OF WASTE

The key elements of simplification are JIT management and product design for better product ability, which both create immediate productivity increases. JIT's work flow organization and the synchronization of the supply chain will reduce the work–in–process. JIT emphasizes the elimination of waste, and waste is all which does not increase the value of the product. Design for product ability requires that design engineers and manufacturing engineers, traditionally adversaries, work as a team to achieve a common goal.

2.2. AUTOMATION

Automation should be introduced selectively and in such a manner that later integration is possible. The bottom–up implementation of CIM isolates islands of automation and their interface and integration problems, and initiates small projects, which produce returns on investment and are part of an overall top–down CIM strategy.

2.3. INTEGRATION

Integration means unifying the parts. There are several dimensions to integration:

- Integration of the functions of a manufacturing enterprise, ie in sales, product development, engineering, supply chain, fabrication, assembly, distribution etc.
- Integration of islands of automation
- Integration of computers and databases

- Integration of mechanical elements, e.g. conveyors,
- AGVs, storage systems, machine tools, robots etc.
- Integration of the control systems applied to the mechanical elements

Increasing number of communication links and networks between systems, equipment, organizations, suppliers and customers will step by step transform manufacturing from islands of automation, into integrated systems. However, the links and networks do not make out CIM. The essential difference is that a real CIM supports the sharing of common data.

3. THE BUILDING BLOCKS OF CIM

The computer and network architecture is the technical backbone of CIM, providing means to execute and monitor the manufacturing system. A good architecture derives from manufacturing requirements, and creates a flexible enough framework for a step by step implementation as well as for future adaptations. Most CIM systems are supposed to be multi–vendor and multi–computer systems with complex data bases and communication capabilities. The two basic types of architectures are called: *distributed or centralized.* Both are based on a network of computers; the key difference is the number of hierarchical layers, which in a distributed system can be up to seven (ISO layers).

The factory network is the backbone to any CIM, providing a standardized linking method between different layers and different computers. The dream of a large transparent networks, where every device can communicate with any other device is far away, but the interfacing of node computers is a practical opportunity today. The linking of data bases and applications is a more demanding task, which requires a lot of customized software.

4. CIM BENEFITS

The typical measurable and expected benefits of CIM are typically the following:

- 30–60% reduction in WIP
- 30–70% reduction in average lead time
- 200–300% gain in capital equipment utilization
- 200–300% product quality gain

The benefits accrue from three major sources, which are the product performance improvements, product cost improvements and indirect benefits. Let's briefly discuss these benefits in this context.

4.1. PRODUCT PERFORMANCE IMPROVEMENTS

CIM will contribute to product quality, since before it can be installed successfully, many long–standing problems have to be resolved. E.g. quality must be built into the process as well as into the product design and specifications. The value of product and product variations can be increased because more detailed analysis can be done with CAD and CAE prior to their manufacturing. The common data base approach gives better and more accurate support to the field service and distribution chain. Technical manuals are easier to keep updated.

4.2. PRODUCT COST IMPROVEMENTS

Process monitoring, feedback and fast rescheduling all contribute to the optimization of manufacturing. The machine utilization rate can be increased and scrap and rework decreased. Due to higher utilization rates and increased effective hours the time required to manufacture the same number of parts and products decreases significantly. The number of machines will decrease, too. Operational flexibility enables rapid product mix changes and repositioning of the marketing. Major improvements can be expected in the amount of overhead work.

4.3. INDIRECT BENEFITS

Better quality, more product variations, the ability to customize, and shorter and more accurate delivery times all increase customer satisfaction and create possibilities to price premiums. The information required by management and accounting can be collected in a natural way. Management accounting could become again what it used to be in late nineteenth century, i.e. management accounting measuring critical operational performances.

5. APPROACHES TO CIM

CIM is still in its infancy. However, certain lessons have been learnt from experimentation. Here is a brief list of some of them:

- CIM is more a concept and a process than a technology issue.
- CIM should be regarded to be a strategic investment program, not just technology acquisition program.
- "Simplify, Integrate, Automate" is a valuable and a very practical advice.
- CIM is not easy. The technology required is not standardized yet, although INTEGRA provides a start.
- CIM covers a broad scope of functions and activities and overlaps several organizational entities.
- Computers are not as fast and not as intelligent as many prefer to believe.

Every manufacturing company has a choice: Firstly, it can wait and see. Secondly, it can install islands of automation". Thirdly, it can apply a "top down – bottom up" approach and try to build systems, which will be relatively easy to integrate now or later.

CIM is emerging and new manufacturing technologies abound both high and low technology. Companies are advised to use common sense and to avoid big technological leaps. There are many low technology opportunities, which on the one hand, increase productivity and, on the other, provide a better foundation for high technology applications.

SECTION II:

TRENDS IN COMPUTER INTEGRATED MANUFACTURING

Chapter 3

The Development of Modern Manufacturing Automation

Professor Hans Andersin
Helsinki University of Technology
Otakaari 1
SF 02150 Espoo
Finland

Abstract: After having defined the terms "Automation" in a wider context, the paper places the INTEGRA initiative along the path of the historical developments of manufacturing automation. It explains the various terms being used today in factory automation. The multitude of new systems, methods, and technologies pose a number of problems, as they have been developed without the intention to work together with other automatic systems. In order to be able to integrate these islands of automation, it is important to work according to a CIM strategy that contains the elements of modularization and standardization.

1. WHAT IS "AUTOMATION"?

The term "Automation" started to be used in the late 1940s in USA. In the English language it has the same meaning as the much older word "automatization" (from Greek word Automatos = moving by itself) denoting that a process of some kind takes place without immediate human control and is able to adapt itself to changed conditions. "Human mental processes carried out by machine" as philosophically inclined persons would say. In some languages, e.g. in Swedish, the word "automation" has a social, political or even sensational nuance to it and is mostly used by laymen, while "automatization" (automatisering in Swedish) is preferred by professionals. Within, or sometimes outside the scope of automation we have terms like "mechanization" representing "primitive" forms of automatic functioning, and "robotization" or "computerization" representing "higher" forms of it.

Throughout this paper we will deal with automated processes–be they manufacturing, chemical, office work, planning, engineering design, transportation, or something else–being capable of functioning at least partly unmanned at least part of the time. We will use the word automation as the collective name of the phenomenon itself, the techniques and methods whereby it is achieved, the theories behind it, and the economic, human, and social consequences of it.

We will not deal lengthily with the earliest roots of automation. The classical example is the steam engine of the eighteenth century representing a mechanical device replacing human or animal muscle power. It uses a centrifugal speed regulator representing an automatic control or feedback function. Another classical example is the mechanized loom of Jacquard of the early nineteenth century that could be programmed to weave different patterns into the cloth. The two techniques represented by the classical examples, feedback control and programming, are still fundamental ingredients of modern automation. Other fundamentals include automatic measurement or sensing in order to be able to observe the state of the process, its environment and its objects, and automatic actuation in order to be able to manipulate the same. Another basic technique is called integration: separate operations are combined, coupled together, grouped or chained to form greater wholes. If we, however, were to define the key technology enabling modern automation we would without hesitation nominate information technology as the chief candidate for this role. Information technology includes, besides computers and Programmed Logic Controllers (PLC), and data–transmission technology, the whole world of software such as database handling, CASE tools, programming systems and languages, expert systems, pattern recognition, presentation systems, network managers, and a myriad of applications. Many of the other key ingredients of automation such as programming, feedback control, and integration can best be implemented by means of information technology.

2. THE EARLY DEVELOPMENT OF MANUFACTURING AUTOMATION

Manufacturing automation was started on the factory floor as an endeavor to increase productivity by replacing human operators by automatic machines. People were seen as slow and unreliable, production rate limiting, quality decreasing, and cost increasing elements in a manufacturing process. Wherever possible, usually in highly repetitive mass production operations, special purpose machines were constructed to produce specific parts or products automatically in large quantities. Productivity increases were manifold and the high investments involved were paid back in a matter of months or a few years. This type of manufacturing automation had its roots in the second world war weapon industry. This development reached maturity in the 1950s. It is still a viable technique whenever the market calls for mass-production and has reached peaks of technological refinement during recent times especially in Japan. The drawback is, besides the high investments required, the inflexibility of the production machinery. Even small changes in product design will require time and money consuming changes or rebuilds of the production machinery.

Another widely used technique, suitable for automatic manufacturing operations, is to use various kinds of automatic machine tools with exchangeable tools, dies, and fixtures. This type of technology was earlier well suited only for long series and large lot sizes. Due to the exchangeability of the tools and fixtures this production technology is much more flexible than the special purpose machinery described above. The drawbacks were long set-up times, high cost of tools, dies and fixtures, high cost of work-in-process, and low machine utilization. As we will see, these drawbacks are now being overcome in a modern manufacturing environment. The main techniques for achieving this are the ones mentioned earlier as fundamental ingredients of modern automation, namely programming, integration, and information technology.

Early programming techniques involved controlling the tool path of an automatic machine tool by means of a physical template or copy of the part to be machined. This technology was obsoleted by numerical control (NC) in the 1950s. In NC the tool paths and other tool control information were punched as a sequence of codes into a paper tape. The tape was read by the paper tape reader of the machine tool and converted into mechanical actions required to produce the part. This technology in turn has been obsoleted by computerized methods to produce the tool control information directly from an electronic drawing of the part. This information can be transmitted directly to the machine tool without the use of paper tape.

Early integration was achieved by arranging the different machine tools required to produce a certain part along an automatic transfer line. The line automatically transferred the work piece from one operation position to the following one. Needless to say this technology requires long series of similar parts in order to be economically and practically justified. The transfer line is still used as a viable concept in some modern manufacturing environments. Other than transfer line arrangements for integrating the shop floor machines are, however, getting more frequent as we will see later.

Table 1.: Some Examples of New Manufacturing Devices and Concepts

1940 COMPUTERS
1950 NUMERICAL CONTROL
1960 COMPUTER AIDED DESIGN PRODUCTION MANAGEMENT SYSTEMS INDUSTRIAL ROBOTS
1970 COMPUTER NUMERICAL CONTROL COMPUTER INTEGRATED MANUFACTURING FLEXIBLE MANUFACTURING SYSTEMS AUTOMATIC GUIDED VEHICLES
1980 FLEXIBLE AUTOMATIC ASSEMBLY SYSTEMS AUTOMATIC STORAGE AND RETRIEVAL SYSTEMS CAD/CAM COMPUTER AIDED PROCESS PLANNING
1990 RAPID PROTOTYPING COMPUTER INTEGRATED ENTERPRISE

3. LATER SHOP FLOOR MACHINERY DEVELOPMENTS

In the 1970s all through the 80s automatic machining centers were developed. Machining centers were obtained by equipping numerically controlled milling machines with built–in tool storage to facilitate automatic tool change. Automatic pallet storage, loaders and unloaders were installed in order to make it possible to preload pallets containing workpieces for several hours of machining. In addition to this many of these machines have multiple functions integrated into them making it possible to turn, drill and mill a work piece within the same machine, and measure the result automatically. To make all this possible the original relatively simplistic numerical controllers on the machine tools had to be replaced by computer based controllers. These are capable of taking care of storing NC–programs, detecting error situations and controlling the sequence of operations. When computers were first used in the controllers the name NC–machine tool was changed to CNC, Computer Numerical Control (today when all automatic machine tool controllers are computer based, the 'C' has been dropped as being redundant).

Computer based numerical control is nowadays used for all kinds of manufacturing machinery and processes. Automatic transportation systems such as automatic guided vehicles, AGV, invented in the 1970s, have reach maturity during the past few years. The same can be said about the large number of Industrial Robots, invented in the 1960s, that are used for picking and placing, loading and unloading, assembly, painting and other applications. Many promising robot applications using automatic vision and other senses are still being developed.

Another important family of computerized shop floor systems consists of various automatic storage and retrieval systems for materials, parts and semifinished and finished products. This type of storage systems are often used as a part of the manufacturing process. The newest addition to computerized numerically controlled manufacturing processes is the stereolithographic process for making parts using a plastic material. The part is "frozen" (polymerized) from a plastic liquid by a computer controlled laser scanning system. This process, connected to a Computer Aided Design system is used for making prototype parts.

4. ENTERS AUTOMATED PLANNING

As was mentioned before, some of the problems resulting from the use of the emerging manufacturing automation technologies were increased levels of work in process and low levels of machine utilization. Other common problems were shortages of critical parts and raw materials just at the moment they were needed on the shop floor. Long and uncertain delivery times of finished products were also a common problem. As these problems were inadequately handled by manual planning means, different mechanized means were introduced. Punched card system for material requirements planning and shop floor control were used already in the 1940s and offered some improvement. In the 1960s when large scale computers started to be used in the manufacturing industry instead of punched card machines, the bold idea of the completely integrated automated materials and operations planning and control system emerged. Using computerized databases containing Bills-of-Materials, timed production operation sequences for every product, and information on each part or material to be procured, the system was to be able to schedule an order, issue purchase suggestions, and prepare shopfloor orders for work and material. These extremely complicated production control systems did not work in a real environment and caused some users big difficulties and losses. Based on these early failures less ambitious and less sophisticated computer based planning and control systems emerged that are widely used today for preliminary planning of manufacturing and of material procurement operations. The detailed planning and control is done more or less manually, often assisted by some computerized tool for visualization or simulation. In order to cope with lead time and inventory problems, however, new organizational measures had to be taken in addition to the computerized planning methods mentioned above. One of the most important of these organizational measures consists of regrouping of the various shop floor production machines according to the part or product to be manufactured rather than according to the kind of operation the machines perform. The new machine groups for parts were called manufacturing cells, or Focused Factories when they contained a large number of different machines for producing a specific product. Another very important measure whereby inventories could be decreased and throughput times drastically cut is called Just–In–Time (JIT) control of the material flow.

This means basically that everything that happens in the factory happens just when it is needed, no later or no earlier. JIT, in order to be efficient, requires that products have similar parts or subassemblies at least at some assembly level. Further the manufacturing system should be capable of making parts with a very short lead time on demand just when they are needed. This in turn requires extremely short set–up times and a highly developed system of material handling. In the early 1970s this demand gave birth to the class of Flexible Machining Systems or Cells (FMS or FMC). FMSs are computer integrated systems consisting of all the various manufacturing machines required to produce a family of parts. The components of an FMS include machining centers and other NC–machine tools, measuring devices, and finishing and washing machines. Material handling means such as computer controlled transporters and robots for moving the pieces between the various units, and automatic storage and retrieval systems are also important parts of an FMS.

Just–In–Time control was soon extended to include outside material suppliers and subcontractors that were required to deliver the materials and parts on demand just in time when these were needed on the shop floor.

5. DESIGN AUTOMATION HAS BEEN WITH US FOR A LONG TIME

Computer Aided Design (CAD) was already in the 1950s used as a name for carrying out the calculations needed for a new product, be it a bridge, a transformer, a transmission or some other nontrivial construction, requiring trial and error adaptation to standardized measures. Upon the advent of relatively high resolution CRT–displays during the 60s and the 70s, the CAD applications grew fast in number, usability, economy, and sophistication. Product database systems for storage and retrieval of drawings, product and part specifications broadened the usability of CAD systems etc. The very drastic decline of computer system prices made CAD even more a standard tool for manufacturing companies. This development was accentuated by the use of PCs (personal or micro computers) for CAD.

An important feature of a growing number of CAD–systems is the ability to generate NC–programs more or less directly from the drawings. The NC–programmer working at a display terminal calls up the drawing of a part, selects a cutting mode and a tool "electronically" by pointing at it and moves it to a place where the cutting should start. The system calculates the tool paths automatically and draws them on the screen simulating the real tool movement. The resulting NC–program is converted to a file of commands for the machine tool controller and is stored for future use. When needed the program is sent to the machine tool by means of a data transmission link. This technique is called DNC, Direct or Distributed Numerical Control. CAM, Computer Aided Manufacturing is used as an umbrella name for computer based methods in manufacturing such as automatic NC–programming and automatic control of machine tools.

A system that includes CAD and CAM in an integrated fashion is called CAD/CAM. One of the latest functions to be automated is the technical production planning that lays out the path of a part through the factory, sequencing the machining and other manufacturing steps, selecting the tools and designing the fixtures, etc. When technical production planning is aided by information technology it is called Computer Aided Production Planning, CAPP. CAPP is still being developed by advanced techniques like features based planning and expert systems. CAPP provides real integration between CAD and CAM resulting in what could be called CAD/CAPP/CAM.

6. THE NERVOUS SYSTEM OF THE MANUFACTURING ENTERPRISE

In DNC the NC–programs are transmitted from the CAD/CAM–system to the various machine tools on the shop floor. At the beginning the information was communicated along individual cables from the central system to each machine tool. At a later stage it was found that a network consisting of one cable common to all the machine tools was sufficient to perform the task required. This resulted in reduced cabling cost and added flexibility. This was the beginning of the era of the Local Area Network (LAN). Modern manufacturing enterprises usually have a number of such LANs connected to each other. The purpose is to facilitate people–to–people communication (electronic mail systems), people–to–machine and machine–to–machine communication. The communications network was extended to encompass suppliers, salesmen, subsidiaries, and customers through private or public communications networks.

Many data transmission standards, some officially approved, some vendor supported "de facto standards" have been introduced. One of the standards called MAP, Manufacturing Automation Protocol, shows some promise to be used for broad band communication on the factory level. A special telecommunication service called EDI, Electronic Data Interchange, is standardized as "EDIFACT" to provide a language to be used for communicating trade documents among different companies and among companies and transportation firms and customs and other authorities. Another set of standards and programs (IGES, PDES, STEP) is used for transmitting product drawings and other product data from one CAD–system to another that may be of an entirely different brand.

7. SIMPLIFY, AUTOMATE, INTEGRATE

The various LANs and other telecommunication networks that we have described above clearly serve the purpose of connecting the various computerized systems of a manufacturing enterprise with each other and with various people groups inside and outside the enterprise. If the intention is to form greater wholes working together towards common goals we call it Computer Integrated Manufacturing, CIM. The term Computer Integrated Manufacturing was coined in the first years of the 1970s by Joseph Harrington defining CIM the endeavor to join CAD, NC–machines, data collection systems, production planning systems, and various shop floor machines together by putting them under common computer control. All of these subsystems were available already in 1973 when Harrington's book, Computer Integrated Manufacturing was published. Later CIM has been extended to encompass all the different systems that we have described above plus purely administrative systems, order handling systems and management systems. The integration is technically accomplished by means of LANs and other networks and by having all systems share data in common databases. Equally important means of integration is material control systems such as Just–In–Time and different organizational means such as cross functional teams. All of this is done in order to save time and money, increase the quality level of the products and services, and the flexibility as to meeting customer needs.

A compact way of stating the purpose of CIM is to say that CIM increases the competitive strength of a manufacturing enterprise. CIM is therefore to be consider as one of the strategies of a manufacturing enterprise. Another reason for calling CIM a strategy rather than a technique or a system is that it calls for a time consuming stepwise development process. This involves people as well as production machinery, management processes and products. A simple way of stating that fact is contained in the slogan: Simplify–Automate–Integrate. This implies that before we start automating or integrating we should simplify our production processes, planning and control procedures, and the organization as far as possible. It is also necessary to simplify the products in order to make them easy to manufacture and assemble.

Automating the individual operations of a manufacturing enterprise often follows the same historical development path as we have described above. Such an evolutionary process may, however, easily result in what we call isolated "islands of automation". Later when time comes for integration of these islands into greater wholes they are found to be incompatible and very difficult to integrate. The hardware and software platforms, used for the different islands may be entirely incompatible. Data may have different formats and values even if they should logically be the same, etc. In order to prevent this to happen and still preserve the step–by–step character of the development process it is necessary to develop a strategic plan, a CIM–strategy or a CIM–plan to be followed when going through the various phases of simplification, automation and integration.

It will also be logical to divide the resulting systems into two classes, the "CIM–applications" and the "CIM–infrastructure". The CIM–applications are carried out by means of the various islands of automation like CAD, FMS, Order Handling etc. The CIM–infrastructure is a name for everything needed to make integration of the CIM–applications successful. The infrastructure includes information technology such as an agreed upon systems architecture, a common logical database, and a network system. It also includes human resources elements such as a trained and motivated workforce, and a number of organizational features for developing CIM, and for operating in a CIM environment. Finally the infrastructure calls for a simplified organizational and operational structure, and for a set of simplified products easy to manufacture. It is essential that the CIM–strategy contains a plan for building up the infrastructure along with the applications in a step by step fashion.

8. SOFT AUTOMATION

Most of the development steps leading to the various islands of automation of a modern manufacturing enterprise uses machinery, computerized systems and other relatively "hard" things. A very important part, however, is always played by a number of "soft" agents such as some of the components of the CIM infrastructure mentioned above.

Many of the soft concepts in modern manufacturing come from Japan where they were implemented earlier and more commonly than in the western hemisphere (although many of them were originally invented in USA). We have mentioned the principle of Just–In–Time before. JIT in Japan is an umbrella concept for a number of things that could be called elimination of waste. Waste of time, waste of material, waste of money, waste of space, and waste of manpower. JIT directly involves elimination of waste of time and work in process and thereby also elimination of waste of material. Another typically Japanese principle is Kai–Zen, the principle of continuous improvement. Development should consist of frequent small steps of improvement initiated by everybody in the organization. Kai–Zen has been called the single most important reason for Japanese superiority over the Western World.

A very important class of soft concepts in modern manufacturing is concerned with different aspects of quality. The most important principles of Total Quality Control (TQC) are the following:

- Everybody in the organization is responsible for the quality within his individual work space.
- The customer's wish and satisfaction is foremost. Everybody considers the next step of the work process as his customer.
- Total Quality means quality in all aspects including products, services, customer and vendor relations, human and public relations, working place orderliness, etc.
- The only number of errors that is allowed is zero.

By means of a technique called QFD, Quality Function Deployment customer requirements are translated into product functions. This technique is widely used in Japan and has recently been employed in western companies as well. Another Japanese quality oriented technique (originally proposed in USA in the 1940s) is called Statistical Process Control, SPC. Some automatic production processes and machines can be set to produce high quality products so that there is a statistical correlation between the machine settings and the produced quality. Samples of the products are measured and analyzed by statistics that indicate if and how much the process settings have to be adjusted in order to produce consistent high quality products. SPC has recently been "reinvented" in the western world.

The concept of Time Based Competition has recently been introduced, again following the example of the Japanese. Everything has to be performed in a fraction of the time it used to be. This concerns lead times for product development and manufacturing, and response time to changes, customer needs, inquiries, and other things. Some companies, e.g. ABB have instituted special programs (the ABB "Time Based Management Program) for cutting all lead times in half and thereby increasing competitiveness and decreasing costs e.g. of capital employed in process inventories and work.

One of the techniques for decreasing the time needed to get a new product from an idea to a production–ready prototype is called Concurrent Engineering. Concurrent Engineering implies that the different engineering tasks are organized to be performed concurrently, overlapping each other timewise, rather than being performed sequentially. Other organizational constructs mentioned earlier are the cell organization and the focused factory layout of the shopfloor. Extending the concurrency principle to the entire organization, using network organizations instead of the hierarchical pyramid, cutting down the number of organizational levels, and introducing self managed groups often go hand in hand with automation in a modern manufacturing enterprise.

Many catchy names have been coined to describe a modern manufacturing enterprise producing highly competitive, high quality products, meeting universal and specific customer needs at less time and lower cost using less resources than its competitors. "World Class Manufacturing" and "Lean Manufacturing" are the best known of these names.

Designing the products to suit automatic assembly and manufacturing is, of course, a very important prerequisite for successful automation. Under the names of DFM, Design for Manufacturing and DFA, Design for Assembly a number of more or less formal methods have been developed.

It is not surprising that an enterprise that have selected the triad "simplify–automate–integrate" as one of its strategies also has to change its way of managing cost rather drastically in order to be successful. This is reflected in the methods used for cost accounting, investment evaluation, and performance measurement. A new discipline called Cost Management Systems, CMS has been emerging alongside with automation in the late 1980s (the main proponents being the Harvard Business School and CAM*I). The three components of CMS are briefly introduced below.

Activity Based Costing or Management, ABC or ABM causes a complete change in the way cost is accounted for in a manufacturing enterprise. Instead of the usual static method of allocating overhead costs to products based on some more or less artificial key like labour or machining hours, costs are primarily assigned to the activities of the enterprise. The enterprise is seen to be a network of activities (operations). Activity costs can be used for various purposes like costing and pricesetting of new products, make or buy decisions, budgeting, and identifying non value added activities.

Justifying investments is most oftenly based on some simple concept like payback time. When investments in modern advanced manufacturing systems compete with investments in conventional technology the payback time principle will usually favour short term conventional machinery over long term advanced systems. We will have to recognize strategic advantages like flexibility, quality, response time, etc. in order to be able to justify investments in advanced systems. Methods and tools for doing this have been developed.

Performance Measurement or Management is a time honoured management concept that has been revived quite recently: if the performance of a person is measured, either concerning his own work and conduct or an activity or a system that he is responsible for, this will influence the person's behaviour and improve the performance. "What you measure is what you get" is a well known slogan indicating the psychological feedback effect resulting from performance measurement. The effect of performance measurement will be further amplified if targets are set for and rewards are associated with the things that are measured. In order for an enterprise to function in an integrated way towards a common set of goals it is important to have the performance measurements and the targets conform to the overall goals of the enterprise.

The development of performance measurement in a manufacturing enterprise started by counting the number of pieces that were made. Soon also measurement of the quality of the pieces was added. The number of work accidents, figures indicating total production volume, and other overall performance indicators were added at an early stage. This was followed by overall indicators like inventory turns, productivity, machine utilization and others. Recently the number of things that are measured has increased considerably: customer and worker satisfaction, total quality aspects, thoughput times, vendor and buyer performance, continuous improvement aspects, marketing performance, etc.

9. CONCLUSION

We have seen how automation has developed in manufacturing enterprises, from some basic techniques like feedback and programming control, through islands of automation like NC– machine tools, Industrial Robots, CAD, and production planning and control systems, to partially integrated systems like Machining Centers, FMS, and CAD/CAPP/CAM.

During the development of these modern manufacturing systems, computer and information technology has been taking a more and more dominant role. This is even more pronounced during the integration phase of the automation development. The islands of automation and the partly integrated systems are connected to form a unit wide or enterprise wide network. The network contains common databases, different manufacturing and design systems, order handling and other business systems.

We have also learnt about the importance of employing "soft" technologies together with the "hard" one in a modern manufacturing enterprise. Reorganizing the shopfloor, Just–In–Time control of material flows, Total Quality Control, Continuous Improvement, Concurrent Engineering, and Cost Management philosophies like ABC and performance measurement are examples of new technologies that are used in a modern manufacturing enterprise.

The multitude of new systems, methods and technologies that are to be used in a modern manufacturing enterprise poses a number of special problems for the enterprise. Many of the systems are a result of evolution from simpler forms and are originally not intended to work together with other automatic systems.

Lack of compatibility is a serious obstacle for integration. This requires converters, interfaces and sometimes extensive reprogramming and database cleaning up as a prerequisite for further development in the direction of CIM. Evolutionary step by step development of the technology through islands of automation is still a recommended way to reach the desired status of being an integrated World Class Manufacturer. In order to be able to integrate these islands it is important to work according to a CIM–strategy or CIM–plan that contains the principles for development and integration of hard and soft parts. The techniques involved are modularization and standardization of interfaces, communications, databases, hardware components, and programs. This involves selection of infrastructural features like a common systems architecture and a common hardware and software platform within which all new development takes place.

REFERENCES

Andersin, Hans, "CIM Strategy Development" in Advances in Computer–Integrated Manufacturing, Vol 1 pp. 3–23, JAI Press Ltd, 1990

Andersin, Hans, "Performance Measurement as an Integrating Link between Man and CIM", Proceedings of the Eight International PROLAMAT Conference, Tokyo 1992, pp 495–502

Brimson, James A, Activity Accounting, John Wiley, 1991

Cole, R. Wade, Introduction to Computing. McGraw–Hill, Inc., 1969

Dreher, Carl, Automation, What it is, how it works, who can use it,. Victor Gollancz Ltd, London 1958,128 s.

Gunn, Thomas G., Manufacturing for Competitive Advantage, Becoming a World Class Manufacturer, Ballinger Publishing, 1987

Kangas, Markku, Computer Integrated Manufacturing – A New Manufacturing Concept, International Management Institute, Geneva, December 1987

Chapter 4

Future Technologies and Standards for Efficient Manufacturing Systems and Software

Louis-François Pau
Digital Equipment Europe
POBox 27,F 06901 Sophia Antipolis, France

Abstract: This Chapter addresses key technologies and standards for manufacturing software and systems, emphasizing the technical aspects, and not addressing the organizational ones. After reviewing the technical challenges, especially those technologies which lead to market dominance, and the difficulties of small lot manufacturing, are surveyed specific critical technology domains: sensors and metrology, machine vision, machine control, fiber optic condition sensors, CIM data networks, data and messaging protocols and standards for manufacturing. Finally trends are given about the design methods applying to manufacturing and CIM software, as discussed elsewhere in this volume.

1. THE TECHNOLOGICAL CHALLENGES IN MANUFACTURING

The challenges can be summarized by four pragmatic observations put forward during the definition of the European FAMOS program, under EUREKA label [10]:

I. towards the mid–eighties, over 40 % of the costs of products manufactured in Europe were due to assembly operations; see [14] for design and analysis hereof

II. at that time, however, European industry was allocating 10 % only of its investments to the automation of specific assembly operations on production lines;

III. automatic cells and even robots were widely used elsewhere

IV. due to high wage levels, flexibility even in small production runs is the key to competitiveness[21].

In addition to this, the Scandinavian environment offers further requirements [9], which are:

V.   the small size of most plants imposing a distributed and concurrent manufacturing concept

VI.   the emphasis on very high quality products and total quality management

VII.   (ISO 9001 [19–20])

VIII.   the pre–eminence of training and reskilling in an attractive way to the workers

It has long been stressed that concurrent engineering at design stage would allow to lower product costs despite similar high labour costs, and that new product lead times were substantially shorter because of the same, and that stock turns can be three times higher because of the use of just–in–time [25]. Also, the creation of multidisciplinary teams (design, manufacturing, logistics, machine design, marketing, information processing) from the product concept to full blown production, is introduced only too slowly to materialize this concurrent engineering in terms of allocation of human assets. At the same time most discussions on total quality are mostly "wishful thinking" unless engineering builds–in so much tolerance and robustness that it can be achieved by shear ruggedizing.

In this Chapter, another view of the future of manufacturing is taken, that is that simple technology and standards [26] are still in the long run the driving forces, as opposed to organizational matters which are intimately related to culture and business goals and therefore hard to spread efficiently. The reason why this approach was little prevailing in the past, is that advanced sensing technology was not thought as reliable, simple to use, and cost–effective in a manufacturing environment, and also that any system would be customized thus denying the standards trends towards open and simpler systems.

The emphasis is thus on emerging information–technology related technologies and standards (listed explicitly) which are considered simple enough to have a deep real impact over the next 10 years for flexible manufacturing meeting the 7 criteria listed above (I–VII.). Section 1 covers sensors and metrology for future CIM systems, Section 2 the CIM data networking aspects, before covering CIM software design and use. References are given.

2. SENSORS AND METROLOGY FOR FUTURE CIM SYSTEMS

2.1. 2-D MACHINE VISION SENSORS

Machine vision has been billed as the prime high speed inspection tool, provided the image processors are fed succinct, unambiguous, directive sensory data. The demand for better quality, tighter tolerances, a better understanding of defects, and more control of the manufacturing process, has shifted the focus from the sensors as monitors of improvement, to sensors as part of the process control that ensures a good part every time.

Machine vision in this sense and role is key to zero-defect goals, as e.g. evidenced in the electronics industry [29]. This means of course that the sensor accuracy must be higher than the tolerances on what is measured. Thus the emphasis is on coupling statistical process control packages, with inspection and gauging systems based on computer vision, all this in real-time environments to meet the production rate flow. This is made possible in particular by linear scan sensors with up to 5000 detectors, and by very high speed frame rate CCD sensors operating on windows up to 128 x 128 pixels. Likewise frame transfer or charge injection cameras without dead areas lead to active vision in moving windows within the field of view. Time delay integration allows the image on the sensor to follow a continuously moving object, instead of requiring more light and a shutter.

Colour cameras and differential colour detection is now possible with CCD colour high resolution detectors, and especially in areas such as food grading and inspection [12].

2.2. IN-SYSTEM COORDINATE MEASURING MACHINES (CMM)

Manufacturing today need dimensional data instantly to avoid making bad parts, and this is therefore a challenge to traditional 3-D CMM's. First, this requires wider use of a standard like DMIS for Dimensional Measuring Interface Standard. Next, non-contact measurement is developing with ruggedized optical heads, scanning laser, and wider range precision stage movements. These systems produce data rate speeds up to 20 kHz, and resolutions to 0.001 inch. New laser interferometric tracking systems can track a movement head in 3-D at speeds of 60 ft/s with accuracies down to 5 microns. This can be further enhanced by angular scanning.

2.3. 3–D MACHINE VISION

The techniques are very varied depending on the application, but include: active triangulation, laser gauges, active lighting, surface finish, 3–D laser scanning (with data rates up to 500 kHz and resolutions down to 0.003 mm), besides projected grid or structured pattern techniques if accurate enough.

2.4. CAD–TO–VISION–TO–NC CONNECTION

CAD directed inspection and sensing is a clear research priority, e.g. in photolitography or in soft metals. The goal is to let the always up–to–date CAD data direct the inspection, with possibly 3–D sensors.

One step further can be reached when flexible configuration tools allow to exploit CAD data to find the geometries such that product customization requires minimum rework, and thus minimum changes to the NC control programs and to the inspection software. Such knowledge based techniques allow for CAD–to–Vision–to–NC code generation [27], e.g. for holes in support beams. This brings to the NC machine shop floor integrated design and manufacturing data, reduces errors due to end product customization, generates NC macrosequences, and reduces measurement work by CMM's.

2.5. METROLOGY

In addition to dimensional metrology by CMM, other major progress is made in metrology. The use of faster processors allows for repeated measurements faster and with better accuracy. In dimensional metrology, interferometric optical chips are slowly introduced with very high accuracy gauging via Moiré or other techniques. In volume and flow measurements, non–contact measurement results from imaging combined with pressure sensors.

2.6. DATA ANALYSIS

The trend is clearly in three directions as related to sensory information:

- creating "world models" to generate the type of nominal or expected sensory information, especially for object positioning or registration oridentification
- learning and adaptation for fast in–process configuration or calibration, as well as to train the system to new defect types, often by doing sensor fusion at the quantitative or symbolic level

- discrete event simulation and control, with hard real–time or dimensional or space constraints, to generate the tasks, their scheduling and possible robot control of handlers, all this from sensory trigger information.

To this end are proposed suitable combinations of spatial data structures, neural networks, image understanding, content based data base retrieval.

2.7. MACHINE CONTROL

More powerful inspection and gauging systems can revolutionize all stages of machine design and control, from model–making to correcting machine parameters when machines or material change in real–time. In other words, one can start by better sensing the machine condition (temperature, forces, and dimensional drift); with this approach, less accurate machines can do tasks that would otherwise require much more expensive equipment.

Microhydraulics is another surprising area, with direct impact on very high speed low power pick–and–place mechanisms (e.g. for robots or transfer lines), as is magnetic levitation for high speed transfer of heavy load.

2.8. MICROMECHANIC ACTUATORS

The big revolution in manufacturing is bound to come in many fields from micromechanic actuators, that is rotating or linear movements from microstructures in silicon, allowing e.g. from microvalves, micro–displacements or alignment, gears, shafts, and ultimately for integrating the PLC controllers with actuators [31]. Take e.g. the realignment of disk drive heads for adjustment or in view of wear.

2.9. FIBER OPTIC MACHINE CONDITION SENSORS

In the future, machine monitoring will therefore rely on a fiber optic nervous system that reports on the machine condition; such sensors can address the previously measurement requirements for temperature (accuracy of 0.1 C), pressure (0.03 kPa), vibrations (1.0 nano–g, rotation rate (0.01 degrees/hr), electromagnetic fields (1.E–12 Telsa), and mechanical stress (0.1 microstrain). What is fantastic is that this can in principle be done with one properly coated fiber. Such sensors are also EMI immune.

3. CIM DATA NETWORKS

Just as the fiber is the nervous system of each machine, the CIM data network is the backbone for the manufacturing process itself and for its coupling with other systems (materials, bill of materials, logistics, spares).

3.1. DATA FLOWS, DISCRETE EVENT QUEUES AND SIMULATION

Information implies some method of transfer, but it must first be known what the sources and sinks are and their order. Therefore, the analysis of data–flow diagrams is paramount to manufacturing control, to show also which information is flowing at any given time (see Figure 4 for an example in inspection in microelectronics).

Data flow analysis reveals how extremely tortuous and divisive (and hence ill equipped for a computer based future) current manufacturing practices are. Efficient technical solutions can only be found in so–called "enterprise integration", where a CIM manufacturing system allows every database to be accessed and updated in a distributed environment, and knowledge to be entered between any two given process stages (see Figure 6) [18]. In this mindset, it is however necessary to carry out a careful analysis of the knowledge persistence and inference tasks for such integration, of which [30] constitutes an early example.

However, this knowledge can rarely be dynamic, and thus discrete event simulation is required covering the data–flow graphs and which encompasses queues and priorities, and allows for a range of possible arrival/work processes ranging from stochastic ones to user–specified event–triggered one's. When these queuing processes are decomposable into the product form, and/or fulfil the "station balance" condition, analytical form results can be used over a range of processes (such as M/M/n–FCFS, M/G/1–PS, M/G/1–LCFS, M/G/inf.–IS) even for closed networks. When no analytical results exist, there are recursive algorithms (vs. nodes and node types) based on convolution to compute the solutions rather efficiently [4–8].

3.2. MANUFACTURING NETWORK PROTOCOLS AND OPEN ARCHITECTURES

One such protocol is MAP, initially promoted by General Motors, and currently in the standardization process [24]; acceptance has been limited by slow conversion to MAP and high costs. MAP is roughly compliant with the 7 layer OSI model for communications, with headers attached to the data frames. To cope with time–critical data transfer two MAP modifications have been defined: MAP/EPA and MiniMAP. MAP is related to TOP, the Technical and office protocol. MAP 2.1 layer specifications are given in Fig. 1 and the details of standards applicable at each layer appear in Fig. 2. The connection to non–MAP systems is via a bridge or router. The connection to the cell controller is sometimes via the IEEE 802 standard interfaces, such as 802.4 (Broadband backbone), 802.3 (Baseband Ethernet), and 802.4 (Carrierband subnet).

The CIM–OSA/AMICE project (ESPRIT project no 688) [11] has been developing an open systems architecture for CIM, to combine MAP, TOP, IGES (Initial graphic exchange specification), and others into an all–embracing conceptual framework, so as to allow CIM users to evolve in an open and decentralized way, and establish migration plans for existing applications. CIM–OSA provides a reference framework with first the CIM–OSA Reference architecture, and next the CIM–OSA Particular architecture, following a building–blocks approach (see Figure 3). Other similar standardization projects or documents include: CAMI DPMM (Discrete Parts manufacturing model), NBS AMRF (Automated manufacturing research facility), ISO TC184/SC5/WG1 N15 (Reference models for manufacturing standards) and some other ESPRIT projects (such as ICAM Project no 1105).

4. CIM SOFTWARE DESIGN AND USE

Production planning and scheduling rely extensively on, first sensor and actuator data, and next on planning goals and resource constraints. While many commercial or custom solutions exist for this, some technical aspects which are still weak in most such implementations, and which are suggested or surveyed here.

4.1. MANUFACTURING PROCESS SOFTWARE VERIFICATION, VALIDATION AND MAINTENANCE

Especially in batch or assembly tasks, it is quite possible to describe formally the complete processes and to explicit the goals/requirements for the process. But it is almost never studied how to provide assurance that the CIM software meets those requirements (which is the verification), nor how to test the CIM software to evaluate its performance and compliance (which is the validation). Verification must be taken into account by the development groups with the purpose of assuring that the software is free of errors, and has the corresponding protections, system resources as required, etc. This verification could find more than inspiration in the ESA (European Space Agency) Software engineering standards PSS–05–0 (1987), ANSI/ANS 10.4.1987.

Validation may be based on statistical acceptance plans and benchmarks, in that formal methods are quite difficult.

Maintenance of manufacturing code can be eased and speeded up substantially by combining reverse engineering with knowledge based maintenance [28], in that the knowledge consists of procedures which will accept or reject a code modification based on consistency with old code, or with a synchronization or communication protocol. The link to reverse engineering is also justified by the reuse of old code in case of a change in O.S.data base, etc.

4.2. REAL–TIME RESOURCE ALLOCATION

Production control in a manufacturing system is defined as the function of directing or regulating the orderly movement of goods and services through the entire cycle. Shop floor control is today relying on off–line resource allocation (materials, equipment, labour) where real–time allocation is necessary. This raises the issue of the design of real–time resource allocation, and one aspect hereof is the lack of constraint based programming languages (such as Prolog III, Charme,..) operating with guaranteed time response. Constraint based programming languages are ideal for the combinatorial search and eventual optimization they perform while pruning the search space through the constraints.

Indicative of this trend in it's functionalities is BASE Star integration software (Digital Equipment), designed to distribute event–driven data collected from a variety of industrial control devices to manufacturing applications; these include Evaluation service, for applicability determination in a given CIM environment, performance and tuning, and integration to heterogeneous systems.

4.3. RAPID PROTOTYPING FOR MANUFACTURING SOFTWARE

Traditionally, manufacturing software has together with administrative software, belonged to the slowest to be developed and especially updated, with a few exceptions. Rapid prototyping [15], with corresponding manufacturing related object and graphic libraries, are bound to be mandatory for flexible manufacturing; this rapid prototyping is bound to push also for formal process specifications (see 3.1), while adding: tools for embedded applications, requirements interpretation, automatic documentation updates.

4.4. DIAGNOSTIC AND TEST SYSTEMS

Diagnostic expert systems can easily be introduced once the machine tools or the whole process are instrumented (see 1.7) and data transmitted (see 2.2) [16,17] (see Figure 5 for an example from a semiconductor process diagnosticsystem). To speed up their development and reduce the knowledge acquisition time, case based reasoning and/or hybrid connectionist expert systems are bound to become necessary. For acceptance and validation, this will however impose some kind of knowledge representation standardization, also for code efficiency; this issue is being addressed by standardization work such as IMKA (IEEE P1272) and AI–ESTATE/ATLAS (IEEE P1024). And the biggest payoff will be through the preventive maintenance function added to the machines.

4.5. HYPERMEDIA FACILITIES

Multimedia facilities will allow a line operator to display on the same console product defect images as well as control graphs or quality control plots. If the machine vision inspection workcell flags a defect, it can store a JPEG (Joint Photographic Exchange group standard) compressed image for on–line viewing (with little network bandwidth utilization) or for post–analysis.

Hypertext is most useful if user manuals and procedures are created within hypertext authoring environments, for faster user adaptation to them.

In a slightly different area, message filtering when applied to the unstructured information collected by manufacturing staff during operations, helps the next shift find this information already organized into folders or structured files.

4.6. COMPUTING ARCHITECTURES AND GROUP COORDINATION

It is becoming clear that distributed computing is winning in the plants, and it is obvious that computing power is not limited by feasibility but by cost and interaction/integration facilities. The issue is then how coordination processes influence the CIM system operations? While the "leitmotiv" for transforming manufacturing processes is either concurrent engineering (see Section 0) or enterprise integration (see Section 2.1), which both place emphasis on technical integration of information islands, the limiting factors may be entirely different. It has been found [22–23] that human coordination processes (including negotiation) limit the amount of effective technological integration, especially as functional managers are unsure or even often disagree on the interactions and technical interconnections. Typical frictions are between CAD, MRP (Materials requirements planning), and computer aided production planning (CAPP); Change Control boards and Scrap review boards overseeing this are found to be often the only places that transcend internal and external organizational boundaries. These coordination processes allow negotiations and flexible handling of external customer constraints and regulations. Therefore it is often unclear how a required multiperson interaction could be substantially improved by computer technology without making he process more inflexible and prone to conflict.

REFERENCES

Journals: Manufacturing engineering (SME), Int. J. Advanced Manufacturing technology (Springer/IFS), Int. J. Computer integrated manufacturing (Taylor & Francis), Industrial metrology (North Holland)

1. K.G. Harding, Sensors for the 90's, Manufacturing engineering, April 1991, 57–61

2. In–Process measurement and control, SME, 1991

3. CIM: proposition pour une mise en oeuvre de la productique, by Siemens group, ISBN:3–8009–1534–0, 1989

4. J.P. Buzen, Computational algorithms for closed queuing networks with exponential servers, Comm. ACM, Vol 16, no 9, 1973, 527–531

5. S.S. Lam, Y.L. Lien, A tree convolution algorithm for the solution of queuing networks, Comm ACM, Vol 26, no 3, 1983, 203–215

6. J. McKenna, Extensions and applications of RECAL algorithm for the solution of queuing networks, Comm. Statist. Stochastic models, Vol 4, no 2, 1988, 235–276

7. M. Reiser, S.S. Lavenberg, Mean value analysis of closed multichain queuing networks, J.ACM, Vol 27, no 2, 1980, 313–322

8. S.C. Bruell, G. Balbo, Computational algorithms for closed queuing networks, Elsevier North Holland, Amsterdam, 1980

9. W.P. Darrow, An international comparison of flexible manufacturing systems technology, Interfaces, Vol 17, no 6, 1987, 86–91

10. FAMOS, EUREKA news, no 12, April 1991, 4–9

11. CIM–OSA/AMICE Program office, 489 Avenue Louise, B 1050 Brussels

12. L.F. Pau (Ed), Fish quality control by computer vision, Marcel Dekker Publ., NY, 1991

13. Proceedings OPTCON'92 (Optical tools for advanced manufacturing), Soc.of Photo–Optical Instrumentation engineers, Boston, 15–20 Nov. 1992

14. G. Boothroyd, Assembly automation and product design, Marcel Dekker Publ., NY, 1991

15. Proc. 3rd Int. Workshop Rapid system prototyping, IEEE Computer society, June 23–25, 1992, Research Triangle Park, NC

16. R. Rowen, Diagnostic systems for manufacturing, AI Expert, Vol 5, no 4, April 1990

17. L.F. Pau, A survey of expert systems for failure diagnosis, test generation and maintenance, Expert Systems J.,Vol 3, no 2, April 1986, 100–111

18. W. Meyer, Expert systems in factory management, Ellis Horwood, NY, 1990

19. ISO 9001, Quality systems: model for quality assurance design/ development, production, installation, servicing

20. J.S. Oakland, Total quality management, Heinemann, Oxford, 1989

21. FMS' 1987, Proc. Workshop on flexible manufacturing systems, Long Beach, CA, 12–15 October 1987, Publ. by SME, Dearborn, MI

22. Y. Bakos, et al, Manufacturing automation as a social coordination problem, Univ. California, Irvine, March 1992

23. R. Kling, Cooperation, coordination and control in computer supported work, Comm. ACM, Vol 34, no 12, Dec. 1991, 83–88

24. G. Nicoletti, CIM: Lan communications, protocol standards and realtime control, SME, Dearborn, MI, 1986

25. A.W. Scheer, CIM–towards the factory of the future, Springer, Berlin, 1991, 2 nd ed.

26. R. Schonberger, World class manufacturing: the lessons of simplicity applied, Free Press, NY, 1986

27. L.F. Pau, D. Paus, T. Stokka, Knowledge based CIM order specific NC drilling system, in: Proc. conf. "Simulation and artificial intelligence in manufacturing", SME, 1988, pp 2–145/2–152

28. L.F. Pau, J. Kristinson, SOFTM: a software maintenance system in Prolog, J. software maintenance: research and practice, Vol 2, no 2, June 1990, 87–111

29. L.F. Pau, Computer vision in electronics manufacturing, Plenum Press, NY, 1989

30. Application of artificial intelligence to manufacturing, Battelle Memorial Institute, Geneva and Columbus, OH, 1987

31. Future view: manufacturing faces the next millennium, Manufacturing engineering, Vol 108, no 1, Jan 1992

32. Design for manufacturability, Vol 6, Tools and manufacturing engineers Handbook, SME, American Technical publishers Ltd, Hitchin, Herts, 1992

Manufacturing Automation at the Crossroads
L.F. Pau and J.-O. Willums (Eds.)
IOS Press, 1993

Chapter 5

CIM-OSA
– A European Development for
Enterprise Integration (1)

K. Kosanke
ESPRIT Consortium AMICE
Stockholmer Str. 7
D–7030 Böblingen, Germany

Abstract: CIMOSA (2), a development carried out by the ESPRIT Consortium AMICE (3), is concerned with the concept of a modelling framework, its application in enterprise modelling for enterprise integration and the IT (4) support for model execution enabling model based enterprise operation control and monitoring. The AMICE project is supported by the European Community and carried out by many European CIM user, vendor, implementor and research organisations.

CIMOSA establishes a framework for enterprise integration. It consists of an open system reference architecture for CIM (CIMOSA) which guides the engineering of enterprise models of particular manufacturing enterprises. The reference architecture guides and supports advanced process modelling of the manufacturing enterprise operations – shop floor operations as well as engineering and administrative processes. CIMOSA supports model engineering and their execution. Engineering and execution of models will be supported by an Integrating Infrastructure developed in the project as well. CIMOSA model engineering places emphasis on model maintenance rather than on one–time model creation. This allows for flexibility in adapting the enterprise operation to changes in both internal and external environments.

CIMOSA enables coexistence with heritage systems through the concept of co–operating processes as well. Non–CIMOSA applications are modelled as independent processes sharing information objects with other CIMOSA or non–CIMOSA processes, are triggering each other via events and exchanging results according to process demands. Integration of non–CIMOSA applications can also be achieved through the use of the functional entity concept. This concept will employ non–CIMOSA application for the execution of a CIMOSA functional operation (lowest level of CIMOSA functionality).

1. INTRODUCTION

Competition in today's world markets (consumer, investment, know–how and capital) is very high and will be increasingly so in the future. To survive and grow in such environments of global competition demands a skilful enterprise management; a management which is able to conform with the continuous changes in customer demands, in technology (manufacturing and supporting IT) and in social and economic conditions. It also requires active cooperation across large networks of vendors and customers in a global economy (5) in order to react in real time on those imposed environmental changes.

To manage change in real time requires both an excellent understanding of the internal and external environment and easy access to the relevant information inside and outside the enterprise. Up–to–date models (6) of the enterprise and its environments and a world–wide industry infrastructure (7) will be solutions for such requirements.

The ESPRIT project AMICE defines and develops an architecture for definition, specification and implementation of Computer Integrated Manufacturing systems – CIMOSA. Such an architecture will become the base for consistent and maintainable enterprise engineering and operation support and thereby be the cornerstone for an industrial infrastructure for industry wide use of information and information technology.

Early project results are summarised in "Open System Architecture for CIM" by Springer [1,2], several publications by project members [3] and a recent private publication by the ESPRIT Consortium AMICE [4]. Standardisation achievements are the European Pre–Norm [5] on the Modelling Framework which also has become the base for the international standardisation in ISO TC 184/TC5/WG1.

The project is supported by the European Community and carried out by the ESPRIT Consortium AMICE which consists of European organisations:

Aerospatiale (F), British Aerospace (UK), Bull (F), CAP Gemini SESA (B), FIAT (I), GRAI (F), Hewlett Packard (F), IBM (D), ITALSIEL (I), ITEM (CH), NLR (NL), Siemens (D), University of Valladolid (E), WZL (D).

2. ENTERPRISE MODELLING REQUIREMENTS

Enterprise modelling has to fulfil a number of user requirements to meet the needs of day–to–day enterprise operation. Modelling has to result in better understanding and handling of complexity in enterprise operation. Modelling has to enable simulation of alternatives and identification of optimum solutions. The ultimate use of a model will be its direct execution to control and monitor enterprise operations. To meet all these needs requires both powerful model description languages and sufficient model creation and execution support. The latter has to support distributed and heterogeneous environments in both the manufacturing and information technology domain.

Models have to describe the operation in terms of functionality and dynamic behaviour (control flow). Functionality should identify inputs required, outputs (results) produced, constraints applied and generated, resources needed and status information created. The model contents has to be presentable in selectable sub–models representing different aspects of the enterprise. However, the model has still to be managed as a whole and changes made at the sub–model level have to be reflected in related parts of the total model. Enterprise functionality and dynamics have to be modelled concurrently and in one and the same model.

Enterprise models have to reflect reality as close as possible and should be used to control and monitor the operation itself. However, different levels of abstraction are required to provide models for decision support in strategic, tactical and operational planning (simulation of alternatives). But all abstractions have to be part of the same model and therefore have to be derived from and linked to the detailed description of the model used for operation execution.

Modelling methods have to provide for evolution in enterprise model building and connection of existing models at a later point in time.

Model engineering has to be heavily IT supported to enable creation of consistent and easily maintainable and extendible models. The user has to be guided through the model engineering process by providing identification of relevant information known to the system already. Ease of maintenance is needed to adopt models in real time to the changing internal and external environment as well as to enable evaluation of alternatives to current reality (As–Is versus To–Be models). The system should also support re–usability and adaptability of existing building blocks (modules).

3. CIMOSA GOALS AND OBJECTIVES

The AMICE project aims on establishing international standards for enterprise modelling and supporting environments for enterprise model engineering and operation. In addition, the ESPRIT Consortium AMICE intends to develop these standards to a level of detail facilitating derivation of compliant product specifications. These will enable consistent business, IT application and physical system integration in heterogeneous manufacturing and information technology environments.

CIMOSA will provide means to:

- develop enterprise models in an evolutionary mode (start with a part rather than the complete enterprise),
- define, describe and structure enterprise requirements in a consistent and meaningful way, derive from those requirements the system design and relevant and sufficient enterprise component specifications,
- describe their implemented versions and support enterprise component implementation and release for operation,
- operate enterprise systems in heterogeneous environments and maintain the enterprise model (adapt to external and internal changes).

Rather than prescribing a single solution to be used by every enterprise, CIMOSA only enables description of particular systems in a consistent and standardised way. Thereby, it enables CIM vendors to develop compliant components and permits CIM users an easy implementation and integration of such components into their particular CIM system.

4. CIMOSA CONTENTS

CIMOSA consist of a Modelling Framework supporting representation of enterprise operation requirements, design and implementation and a supporting distributed TT environment the Integrating Infrastructure. The latter aims on model engineering and execution for enterprise operation control and monitoring.

The modelling framework consists of a reference architecture and a particular architecture which describes the structure of a particular enterprise. The contents of the reference architecture is used to engineer models of enterprise operations.

The reference architecture (generic and partial level) provides the set of constructs which allow to model the operation of a particular enterprise. Generic constructs (Generic Building Blocks and Building Block Types) are applicable to all industrial enterprises whereas more specific macro constructs (Partial Models) are aimed at modelling specific enterprise domains and special industries. These building blocks and macros are used to create and maintain the enterprise models.

Due to their specific interests, users like to focus on different aspects of the model rather than to look at the model as a whole. Therefore, CIMOSA provides four different views (Function, Information, Resource and Organisation View) which allow to analyse and optimise a particular aspect of the enterprise operation without having to take into account all the other aspects at the same time. However, all views are part of one and the same enterprise model rather than being independent models.

Modelling support is aimed on the business user rather than information technology (IT) professionals. Therefore, emphasis is placed on providing means for expressing the contents of a particular enterprise model in a user related language rather than in IT terminology. Relations to IT descriptions are provided for system design and implementation.

The CIMOSA reference architecture supports the different phases of a CIM system life cycle. It starts with modelling of system requirements (Requirements Definition Model). From these requirements the system design (Design Specification Model) is derived leading through system build and release (buy or build, installation, verification and qualification) to the description of the implemented CIM system (Implementation Description Model).

The released version of this model will be used for enterprise operation monitoring and control. Changes will be implemented in the relevant engineering phase and an updated version will be released for operation.

In addition to the modelling concepts, the AMICE project develops an Integrating Infrastructure which supports model engineering and model execution in heterogeneous manufacturing and information technology environments. This integrating infrastructure provides a set of generic services which support model generation and execution. These services will be provided by information technology but will support operation of manufacturing technology through operation control and monitoring (business entity), information supply and removal (information entity), standardised connections to terminals, machines and applications (presentation entity), system distribution and network control (common entity) and overall system management (system entity).

All these services will be part of the enterprise system and as such will be recognised in the enterprise model as IT resources (functional entities) used to support model engineering and model execution.

The Integrating Infrastructure makes use of existing and emerging IT standards (OSI, MAP, TOP, ODP, etc.) as much as possible. It is not the intention of this work to develop competing standards but to complement and enhance established and emerging standards.

Architecture is understood by CIMOSA as:

- CIMOSA Particular Architecture: the structure and contents of a particular enterprise model.
- CIMOSA Reference Architecture: the collection of constructs which allow to structure and engineer a particular enterprise model.

5. APPLYING CIMOSA – VISION IN ENTERPRISE MODELLING

CIMOSA can be used and applied in enterprise modelling in two complementary ways:

- Enterprise Level: to model enterprise operation as a set of cooperating (domain) processes (CIMOSA and non–CIMOSA).
- Process Level: to model CIMOSA domain processes as a network of functions connected by an explicit control flow.

At enterprise level the concept of domain processes allows to model the enterprise as a set of loosely coupled processes sharing common enterprise objects and communicating with each other via exchange of object views and events. Events may be carried by object views (e.g. a plan, an order, a part) or may be generated by a calendar function (due date). Process results will be used by domain processes internal or external to the enterprise. Such domain processes may either have an internal structure completely modelled according to CIMOSA or may be non–CIMOSA applications with a foreign internal structure. Domain processes are supposed to represent rather large areas of an enterprise functionality which produce highly aggregated results (parts machining, product assembly, product development, product engineering, production planning, business administration are potential enterprise domains).

CIMOSA supports coexistence with heritage systems through the concept of cooperating processes. Replacing ancient applications with modern domain processes built according to CIMOSA enables enterprises to evolve gracefully from their current heterogeneity in IT applications into a more and more unified and standardised environment.

At process level CIMOSA provides means for describing lower levels of functionality and the related dynamic behaviour or flow of control inside the domain process. Domain and business processes represent behaviour and enterprise activities functionality of the domain process.

The result of such a recursive decomposition is an executable network of enterprise activities. The enterprise activities are connected through procedural rules (PR's) which describe the control flow (the conditions for selecting the next enterprise activity or activities after ending the previous one). This control flow for the different levels of decomposition together with definitions of domain process triggering events is the contents of the different processes (DP and BP). Therefore, the decomposition itself leads to a nested set of procedural rules pointing to executable enterprise activities.

Each enterprise activity identifies the inputs for and the outputs of its particular function, the required resources and any constraints/controls on the function itself. In the system design process these enterprise activities will be further decomposed into a functional operation network. The latter will be directly executed by enterprise resources.

To describe available resources CIMOSA provides the Functional Entity construct. These constructs recognise resources through its information processing capabilities (send, receive, process, store information). Examples: Human beings, manufacturing and information technology equipment and software modules capable of information processing as defined above.

The functional entity relates to the process functionality (functional operation) at the system design modelling level. The CIMOSA Integrated Infrastructure itself can be modelled as a set of functional entities. The Concept of functional entity can also be used for integration of non–CIMOSA applications by employing these applications to CIMOSA functional operations.

Different methodology can be applied to create CIMOSA particular models which capture the reality of an enterprise. CIMOSA allows for different approaches including top–down and bottom–up. The top–down approach starting with enterprise system requirement definition, continuing with design specification and ending with a description of the implemented system supports traditional life cycle models.

However, a bottom–up methodology may be more suited to describe existing systems. Starting with the identification of functional entities in the particular enterprise domain will lead to the definition of functional operations executed by those functional entities. Aggregation of functional operations into enterprise activities and identification of the related control flow will lead to the structure of business and domain processes. Other methodologies may be applied. However, within any methodology iterations with previous steps of the modelling process (either top–down or bottom–up) will be required.

6. PROJECT RESULTS

AMICE has developed the CIMOSA concepts under previous ESPRIT contracts (ESPRIT I, II and III) during the years 1985–1992(8), resulting in publication of the CIMOSA technical specifications, CIMOSA Technical Baseline [4](9). Its current contract under ESPRIT II (EP 7110) is concerned with implementations of these specifications and covers the time till March 1994.

With the CIMOSA concepts being generally applicable to many types of business, AMICE has focused on the core of manufacturing enterprises: product development, production planning and manufacturing.

The project has developed the framework for enterprise modelling and has achieved significant recognition and acceptance for this framework in the public domain. The basic contents of the Reference Architecture (generic building blocks) bas been developed for the major part of the requirements modelling level. Other parts of the framework have been populated as well and work on building block types has been started to provide more details of the user language for enterprise modelling.

Validation of the basis modelling concepts has been performed through case studies in different industries represented in the consortium (Aerospace, Automotive, Electronic) as well as in different areas of manufacturing (product development, production planning, parts production). Cooperation with other ESPRIT projects (CIMPRES, CODE, VOICE) extends validation into the Machine Tool and Process industry. These case studies are carried out in close cooperation with the AMICE project. Focus is on production control in machine tool manufacturing, automotive assembly, aluminium casting. Additionally, some partial models will be developed in the area of machine tool manufacturing and enterprise information models.

Development of the Integrating Infrastructure has resulted in a basic structure of the set of generic services. Definition of their functionality and relations with each other have been defined as well. Mapping to existing standards (ODP, OSI) is currently done to identify requirements for alignment with and adaptation of currently available or emerging standards. Validation work is also carried out in cooperation with the VOICE project.

7. CONTRIBUTIONS TO STANDARDISATION

Results from project work has lead to a European Pre–norm (ENV 40 003) [4]. This ENV is also the base for the current ISO standardisation activities on architecture. Work on modelling constructs based on AMICE project work has been started in national standardisation committees. Inputs to standardisation work on infrastructures will be provided as well. The need for harmonisation with CIMOSA has been recognised in ODP and STEP.

8. PROJECT STATUS

Emphasis of current work is on pilot implementation of both model engineering and model based operation control and monitoring. Implementation sites are with FIAT and WZL. This work will provide additional validation of the current set of technical specification. Harmonisation with other standards (OSI and ODP in particular) and demonstration of overall capabilities and benefits of CIMOSA to the industry are the major goal of the implementation efforts.

9. STATE OF THE ART

In the area of enterprise modelling the state of the art is either on specific enterprise models(10), or on high level reference models [6]. Only a limited number of current projects center on the enterprise as a whole (EIP and CAM–I/CIE in the US; ISA and AMICE in ESPRIT). Most of the complementary work is focused on particular aspects of the enterprise ranging from product engineering (DICE), to production operation in particular industries (SEMATEC) and logistic support (CALS).

An overview on these activities is provided in the footnote (US based activities(11) and ESPRIT projects (12)). This work can be categorised in terms of the AMICE work relating to efforts on Modelling Frameworks (13) and enterprise modelling (14).

More related state of the art exists in IT application integration. Beside proprietary platforms of IT vendors multi–vendor solutions are reported in both US and ESPRIT projects (15).

NOTES:

1) This is an updated version of a paper presented at the ICEIMT in 1992 [3].
2) CIMOSA = Computer Integrated Manufacturing – Open System Architecture.
3) AMICE = European Computer Integrated Manufacturing Architecture (in reverse).
4) IT = Information Technology.
5) CALS (Computer Aided Acquisition and Logistic Support) an example of a current solution for this requirement.
6) IDEF (ICAM Definition with IDEFO, I and 2) a starting point to fulfil modelling requirements. However, more powerful modelling languages are required to achieve consistent and expandable model engineering, ease model maintenance and provide support for real time enterprise operation.
7) The success of the automobile as the major means of transportation was due to the establishment of a sufficient infrastructure (roads and network of gas stations) and their accompanying traffic rules, regulations and standardisations (location of steering wheel, gas and accelerator panel, etc.)
8) ESPRIT Projects 688, 2422 and 5288
9) The CIMOSA technical specifications have been published by the ESPRIT Consortium AMICE and may be obtained from the AMICE Office, Bessenveldstraat 19, B–1831 Diegem/Belgium
10) ICAM, CAM–I, others
11) US based activities:
    CALS–CITIS: Computer aided Acquisition, Logistics and Support CALS Contractor Integrated Technical Information Services
    CAM–I/CIE: Computer Aided Manufacturing – International/Computer Integrated Enterprise
    CFI: CAD Framework Initiative
    DICE: DARPA (Defense Advanced Research Projects Agency) Initiative in Concurrent Engineering
    IDEF U.G.: ICAM (Integrated Computer Aided Manufacturing) DEF inition User Group

EIS: Engineering Information System
EIP: Enterprise Integration Program
IISS: Integrated Information Support System
IPO: IGES/PI:)ES Organisation
MKS: Manufacturing Knowledge System
PDES IDC.; Product Data Interchanging using STEP
RAMP: Rapid Acquisition of Manufacturing Parts
SEMATEC: Semiconductor Manufacturing Technology

12) ESPRIT projects:
CAD*I (EP322) CAD Interfaces
CIM PLATO (EP2202) CIM system Planing Toolbox
CMSO (EP2277) CIM for Multi Supplier Organisation
CNMA (EP26t7) Communication Network for Manufacturing Applications
IMPACS (EP2338) Integrated Manufacturing Planning And Control System
IMPPACT (EP2165) Integrated Modelling of Products and Processes using Advanced Computer Technology
ISA (EP2167) Integrated Systems Architecture

13) US: EIP, CAM–I/CIE, DICE, EIS, CFI, SEMATEC, CALS
ESPRIT: ISA, AMICE

14) US: EIP, EIS, CFI, (PDES Inc.), IISS, SEMATEC, (CALS)
ESPRIT: ISA, AMICE, (CAD*I), CIM PLATO, IMPACS, (CNMA), CMSO

15) US: CAM–I/CIE, Dice, MKS, EIS, CFI, PDES Inc, SEMATEC, RAMP
ESPRIT: ISA, AMICE, IMPPACT, CIM PLATO, IMPPACT, CMSO

REFERENCES:

[1] ESPRIT Consortium AMICE, Open Systems Architecture for CIM, Springer 1989
[2] ESPRIT Consortium AMICE, CIMOSA Open Systems Architecture for CIM, Springer 1993
[3] selected AMICE publications 1990–1992:
3 papers on CIM–OSA in International Journal of CIM, pp. 144–188, Vol 3, 1990;
3 papers on CIM–OSA in Computing & Control Engineering Journal, pp 101–125, Vol 2, 1991;
4 papers on CIMOSA in Proceedings of ICEIMT 1992, pp 179–225, MIT Press 1992.

[4] ESPRIT Consortium AMICE, CIM–OSA Open System Architecture for CIM, Technical Baseline (CIMOSA technical specifications), private publication 1993

[5] CEN/CENELEC ENV 40 003: Computer Integrated Manufacturing Systems Architecture Framework for Modelling

[6] Williams, T. J., A Reference Model for Computer Integrated Manufacturing from the Viewpoint of Industrial Automation, Proceedings 11th IFAC World Congress, Tallinn, 1990

Manufacturing Automation at the Crossroads
L.F. Pau and J.-O. Willums (Eds.)
IOS Press, 1993

Chapter 6

FUTURE DIRECTIONS IN FMS:
Object Oriented High Level Petri Nets for Real–Time Scheduling of FMS

Antonio Camurri, Paolo Franchi, and Maurizio Vitale

DIST – Department of Communication, Computer and System Sciences
University of Genoa, Via Opera Pia 11A, I–16145 Genoa, Italy
e–mail: {music,pan,foobar}@dist.unige.it}

Key words: High Level Petri nets, Timed Petri nets,
Concurrent scheduling,
Flexible Manufacturing Systems, CASE.

Abstract: This paper introduces a class of High–Level Petri Nets, called tau–nets, used for the analysis, description, and sub–optimal solution of a general class of problems of process scheduling. Tau–nets include timed transitions, a type–checking mechanism, and the definition of taxonomic hierarchies of token types. Objects are typed, including transitions, places, tokens, and the (sub)nets that can be built combining those elements. Furthermore, transitions and tokens belong to two disjoint hierarchies of types supporting a single inheritance mechanism. A class of problems in the FMS area is taken as a reference scenario in the paper. In particular, the general class of problems is characterized as follows: there is a set of concurrent processes (modelled as colours in tau–nets), each formed by a number of temporally related tasks, segments (transitions). Tasks are executable by different resource sets (sets of places), different in both performance and costs. Processes and tasks are characterized by release times, due dates and deadlines. Time constraints are also present in the availability of each resource in resource sets. It has been proved that such a problem does not admit an algorithm for finding the optimal solution in a polynomial time. Our model allows to describe, analyze, and manage problems of this class, and includes a distributed algorithm able to find a sub–optimal schedule according to a set of optimization criteria, based on tasks and processes times (earliness, tardiness) or time independent costs of resources.

The paper focuses on the automatic synthesis of Petri nets models of the coordination subsystem, starting from the problem knowledge base; the dynamic behaviour of the coordination subsystem, whose kernel is a High Level Petri nets executor, a coordination process based on an original, general purpose algorithm. The approach described in this paper is at the basis of a joint project with industrial partners for the development of a sophisticated CASE tool for the simulation of blast furnaces.

1. INTRODUCTION

In this paper we propose an approach for the sub–optimal solution of a particular class of problems of process scheduling, which can be characterized as follows: there is a set of concurrent processes, each formed by a number of temporally related tasks (segments). Tasks are executable by different resource sets, different in both performance and costs. Processes and tasks are characterized by release times, due dates and deadlines. Time constraints are also present in the availability of each resource in resource sets. It has been proved that such a problem does not admit an algorithm for an optimal solution in a polynomial time [1]. The proposed algorithm finds a sub-optimal schedule according to a set of optimization criteria, based on tasks and processes times (earliness, tardiness) or time independent resources costs. The problem class faced in our approach includes both multiple processes and time constraints.

Process scheduling is a general problem [2][3]: situations in which such a problem arise can be found in many research fields as well as in practical applications. Communication networks dealing with packet switching should find a possibly optimum routing policy in order to minimize queue size and time delays. Hard real–time systems [4] also show the need of efficient strategies since they deal with precedence relations between tasks. In operating system design, a great deal of attention has been devoted to the scheduling problem [5]; time constraints are also important to synchronize the access to shared resources, such as data or I/O devices.

Such a problem may also be found in decision support systems [1], where it is known as the project management problem. Process scheduling finds useful application also in non classical domains. For example, computer music performance systems [6][7] are ``computer systems connected to input devices (including musical keyboards or other real–time instruments) and to graphics and audio output devices'' [8], requiring real time scheduling strategies for the management of events coming from input devices and of actions to be performed.

Xu and Parnas [9] recently developed a preemptive scheduler for multitask applications on a single processor.

Our approach starts from a different hypothesis: it allows the scheduling among a number of processors but it is not preemptive, as in most industrial machines.

The proposed architecture is shown in figure 1,

Figure 1: System overall architecture

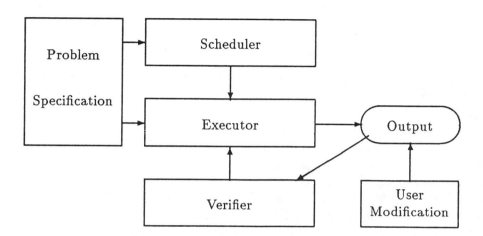

which clearly distinguishes the mechanism of net execution from the special purpose decision support system.

Basing on this fact, two conceptually distinct levels (corresponding to two different, interacting modules) have been developed: the executor (i.e. the coordination process), and the real–time scheduler.

The executor is a fast, efficient tau–nets animator, without any special–purpose problem–solving capabilities to handle conflicts. The scheduler, on the other hand, is the adaptive, distributed component, whose structure and behaviour may heavily depend on the problem class, even if a general architecture can be adopted in a number of possible cases. A default scheduler is always available to solve conflicts when the user fails to provide a specialized one.

Section 2 gives an overview of the FMS problem class, consisting of a production line in which different types of workpieces must be routed into a network of tool machines, trying to satisfy user defined optimization criteria. Section 3 discusses the use of PNs in the modelling of FMS. Section 4 introduces the tau–net model and presents the basic features of system kernel. The decision situations which are demanded to the real–time scheduler subsystem are described in section 5. The definition of a proper high–level representation environment for this class of problems is briefly explained in section 6.

The methods for automatically deriving a model of the coordination subsystem of the FMS in terms of PNs are also described. In [10] the reader can find major details on the structure of the FMS knowledge base.

2. PROCESS SCHEDULING: THE CASE STUDY

The case study has its primary focus in the scheduling of concurrent projects, where a project is a set of temporally related tasks, whose performance gives a complete production cycle of a workpiece in a FMS. Basic terminology and some definitions are borrowed from [1].

Different projects correspond to different production cycles for different workpieces and may be conflicting in the use of machines and resources. In general, different pallets will be needed for each workpiece type in the FMS. Therefore, the number of different pallets available in the FMS defines the maximum number of projects which can be carried out. A set of projects has to be run simultaneously, that is, different types of workpieces must simultaneously be processed by the FMS. Tasks activities are defined in terms of functions.

A task can be assigned to one or more machines (stations) in the production line.

Machines are a class of possible resources; in general, a task can be accomplished by different resource sets. The decision on the resource set for a given task instance at a given time is demanded to the real–time scheduling subsystem, which operates according to the optimization criteria and the constraints defined by the problem. Each resource is characterized by an availability table, which specifies the time intervals in which the resource is available: this models real world situations, such as periodic maintenance of machines, or possible periods of time in which a machine is devoted to other higher–priority activities, external to the actual model.

The architecture of the FMS should allow the continuity in the production process, even in case of unavailability of machines whether programmed or not. Tasks in a given project are linked by temporal relations. In general, temporal relations specify either simple precedence relations, or more complicated explicit time relations. Waiting times can be defined between the ending time of one task and the beginning of a subsequent one. More complex definitions can describe the overlap of tasks.

In a FMS domain, this latter case is generally unnecessary, since a workpiece is generally processed by a single task at a time. In any case, our approach allows general temporal relations among tasks to be defined [11]. Waiting times between two sequential tasks may have soft constraints: for example, after a painting task it may be necessary to wait for a variable amount of time, before allowing the subsequent task to start. This is another typical decision situation: the coordination subsystem demands all decisions to the real–time scheduling subsystem, which should determine the optimum value for each case, according to specified criteria.

Figure 2: A production line

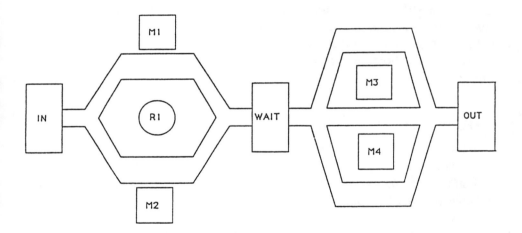

In figure 2 a two projects example of FMS scheme is shown. Each project consists of a sequence of two tasks for a different workpiece. For both projects, the first task can be accomplished using either machine M1 or machine M2; in both cases a further resource R1 is needed. Therefore, Task 1 has two possible Task Instances (TI), named TI11 and TI12, corresponding to the use of M1 or M2.

The same situation occurs for Task 2, with the only difference that in this case we have three TIs (TI21, TI22, TI23), and machines M3 and M4 are needed from TI22, so they appear in two different resource sets. We assume that each resource in the example (generic or machine type) has an availability table, consisting of a set of couples of time points, defining the time intervals in which such resource is available.

Yet, a waiting time is needed between the end of task 1 and the start of task 2, for both projects. Each project has different waiting time ranges. A production line in which different projects are concurrently processed needs a scheduling policy to optimize the productivity, basing on criteria and costs of some kind. We assume that tasks and projects are constrained by their workpieces release time, due date and deadline, which are the initial time at which the task or project could start, the due time for its completion, and the maximum time allowed for its completion. Optimization criteria, based on task and project times, or on time independent machine costs, are taken into account in the problem formulation (see section 5).

System properties, both temporal and structural, for the definition of the problem are represented in a FMS knowledge base, implemented as a structured frame–based multiple inheritance semantic network, described in detail in [10]. A short overview of the representation scheme can be found in section 6.

3. MODELLING FMS BY MEANS OF PETRI NETS

Petri nets find useful application in the modelling of systems characterized by a distributed and concurrent nature, by the synchronization among tasks in the use of shared resources [12][13][14].

Nevertheless, PN–based models of real systems are difficult to manage, since the rapid growth in size and complexity of the net. Recent research in software engineering and Petri net theory proposes integrated approaches to increase the modelling power of PNs. New formal extensions to the original PN model (P–T nets) play a fundamental role in the development of formalisms for visual executable specifications and rapid prototyping.

Timed PNs and Stochastic PNs [12], Coloured PNs (CP–Nets) and Predicate/Transition nets (Pr–T nets) [15] are among the most significant extensions, and are at the basis of several CASE environments [16].

High–Level Petri Nets (HLPNs) are a powerful tool for the definition of models of concurrent distributed systems at higher level of abstraction. This fact is of considerable help in the modelling of systems consisting of large numbers of resources and processes. HLPNs allow the definition of individual tokens, and transitions behave in different ways according to the firing colour (selected from the input places): a transition in a HLPN corresponds to a set (a class) of transitions in the P–T nets model.

This can be considered a particular object oriented approach, in which classes of tasks, corresponding to a transition, are connected to their related resource sets (set of places). Different approaches have been introduced for introducing object–oriented concepts into PN theory: for example in [17] objects are defined as subnets (with a fixed topology) in an extended P–T net model (PROT nets); high level transitions are objects representing subnets; in [18] a software development methodology that uses SADT diagrams and Hierarchical CP–Nets is described.

Several other approaches for the modelling of FMS using PNs have been recently developed: in [19], PNs are used in conjunction with knowledge based paradigms (rule–based systems) for monitoring tasks of FMS; in [20], Stochastic PNs are introduced for the performance evaluation of FMS; yet, in [21], the problem of designing the control system of a FMS basing on Coloured PNs and rule–based techniques is faced; a control system for FMS based on both an extension of Coloured PNs and artificial intelligence concepts (yet rule–based systems) is presented in [22]; the integration of PNs and Operation Research methods in an Expert Systems environment is proposed in [12].

Finally, the hybrid A–Nets model is introduced in [23], for high level representation of robotic actions. A–Nets combine a structured frame–based knowledge representation with HLPNs.

4. THE TAU–NETS MODEL

Before choosing a PN model, a simple but relevant consideration should be kept into account: the more powerful the model, the more difficult (less tractable) computationally it becomes.

Therefore, if the basic goal is analysis, then the minimally sophisticated PN model should be chosen. On the contrary, for simulation–based applications (as in our case), based on an ``analysis by synthesis" method, the choice of powerful PN models can be advantageous. The basic features of the tau–net model are here briefly discussed. We adopt an extension of Coloured PNs, including timed transitions, and hierarchies of tokens and nodes. Tau–nets are a natural extension of STCPNs (Structured Timed Coloured Petri Nets) [24].

4.1 AN INFORMAL INTRODUCTION TO TAU–NETS

Colours correspond in our model to the different projects: at any given time, a project state is identified in a tau–nets by the set of tokens of its colour. The flow of tokens of a given colour in the tau–nets model represents the evolution of the associated project. A neutral colour is defined. A resource (place) marked with a neutral colour can be utilized by any project, that is, a neutral coloured token can be used in any coloured firing occurrence.

Transitions correspond either to TIs (TI–transitions) or to waiting processes (W–transitions), and are characterized by a processing time, which can be deterministic or stochastic. Immediate transitions (with a null processing time) are allowed; their default management in case of conflicts follows [25], that is, conflicts between an immediate and timed transitions are always solved in favour of immediate transitions. Of course, loops and sequences of immediate transitions are not allowed.

The information on the processing time of a TI–transition is particularly useful in case of system simulation. In a real world environment we have real (low level) local controllers, each of them `linked' with a suitable TI–transition at tau–nets level: the firing of a TI–transition causes messages to be sent to its controller for the activation of the corresponding real task instance; the expiring of the transition can be driven by the completion of the task instance, via messages coming to the transition from its controller. In a simulated environment, we know the parameters of the machine(s) executing the task (for example the speed) associated to a TI–transition, and we can use them for computing its processing time, either deterministic or stochastic. The function associated to a TI–transition can be invoked at any time during the transition activity: typically, it is called on transition firing. Other frequent cases are the expiring time (i.e., when the output places are marked), and each time slot during transition activity (see below, the method tick() for a transition).

W–Transitions may have soft–constrained processing times: only the range is known. The real–time scheduling subsystem will decide a suitable value in the defined range for each soft–constrained W–transition. Other basic properties of transitions are a status, the set of the allowed firing colours, and the function defining the code for the local controllers; for each transition, a different behaviour can be defined for each allowed colour (i.e. project).

Places can be of two types, corresponding to either resources or queues. Graphically, resources are circles, queues are hexagons. A place can have a set of tokens, possibly of different colours. A place of type resource is characterized also by a possible availability table. The availability table defines the set of time intervals (couples of time instants) in which the resource is available. A resource set is defined for each TI–transition as the set of all its input places of type resource. Different resource sets can have places in common: this means that there are topological conflicts in the PN. For example, in figure 4, the resource set of TI11 is the set of places formed by M1 and R1 (I_Queue1 is a queue place). The resource sets of both TI11 and TI12 contain the resource R1: this corresponds to the topological conflict between the two transitions.

The firing rule is the same adopted in CPNs, with the exception of the delays due to the transition processing time. In fact, a transition firing causes the consuming of input tokens, and only after the transition processing time has expired, the output tokens are placed on the output places. This causes tokens to appear/disappear in the net: they are stored in transitions for the duration of their processing time. This is a bad feature, from the point of view of analysis properties: it is possible to reconduce this behaviour to the standard CPN model if we consider a (timed) transition as a particular subnet, simply formed by a sequence of transition–place–transition (see figure 3). These transitions are immediate: the first one consumes input tokens, and marks the intermediate place; after the processing time, the final transition fires (it is the firing rule which changes, now), consumes the token from the intermediate place and produces output tokens. Therefore, any timed transition in our PN model can be seen as a sequence transition–place–transition.

Figure 3: The model of a transition in tau–nets

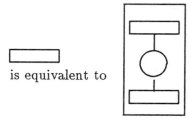

is equivalent to

As for the firing rule, a transition fires if all input places are marked with a given colour, or some places are marked with the neutral colour and the remaining places with the same colour. The firing colour must belong to the set of the allowed colours for the transition. Multiple firing of the same transition, due to the availability of multiple pre–sets, is generally not used in this model, even if it is allowed by the HLPN model.

A further condition has been added to the firing rule of transitions, because of the presence of availability tables on resources: an input place is available if it is marked and at the current time, and within a `reasonable' time horizon, the resource is available from its availability table. The `reasonable' horizon is deterministic in the case of FMS simulation: it is the time point obtained adding the current time to the processing time (computed for the firing colour) of that transition.

Figure 4: The tau−net for the FMS case study

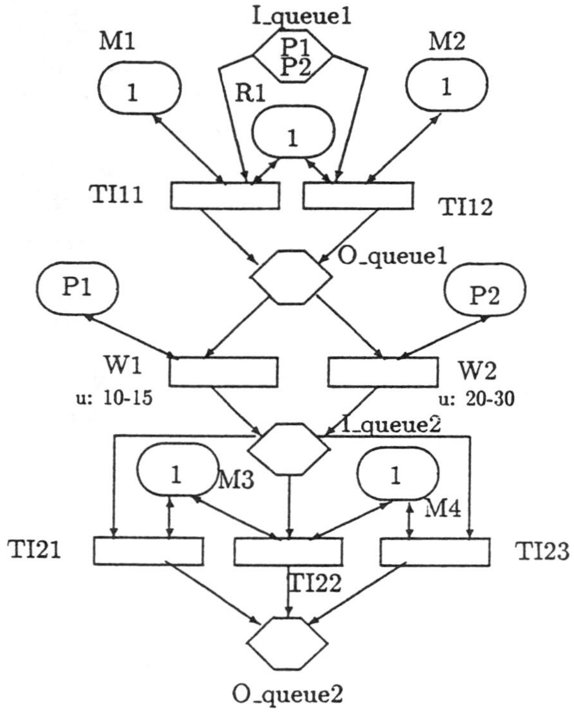

In figure 4 the net related to the FMS case study is shown. Note that this net could be collapsed into a smaller tau−net, using the properties of HLPNs: TI11 and TI12 could be collapsed into a single transition, as well as TI21, TI22 and TI23, with a slightly different firing rule, allowing multiple concurrent firing for disjoint pre−sets [26]. We have chosen in this case a representation with separate transitions, because of its major resemblance to the problem nature, its easier understanding and better tractability. In other cases we use multiple firing transitions.

4.2 TAU–NETS: OBJECT ORIENTED FEATURES

In an object oriented framework it is possible to abstract the fundamental properties of the different components of a HLPN in a natural way.

Net objects belong to type hierarchies which serve the dual purpose of abstracting out commonality in behaviour, thus enabling code reusal, and permitting a formal verification of some run–time properties of the net, by means of a static type checking phase.

These two goals are achieved by means of two distinct hierarchies. The first one, derived from the base class Token has an extension to which every token in the net belongs.

In order to build a Coloured Petri Net we have subclassed Token to obtain ColouredToken, a class of tokens having a slot representing the colour value for its instances. This hierarchy can be incrementally augmented to describe the objects needed in real applications. Tokens are used as argument to the code that implement the transitions behaviour, and the user is often able to trade off simplicity with the ability to find errors during type checking. When programming in the large, it can be useful to subclass tokens even if no slots are added just to get a more strict type checking.

The other hierarchy, made of classes whose instances are the transitions present in the net, has the main purpose of abstracting out uniformity of behaviour and allowing code sharing. The elements in the hierarchy reflects the structure of transitions, as described in the previous section.

Figure 5: The transition hierarchy (excerpt)

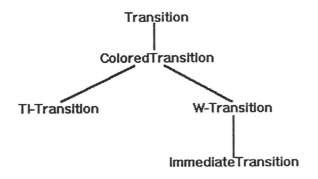

The base class Transition defines various methods, some of which are virtual, and need further refinements in the derived classes. The definition of this class is given using a C++ like notation:

```
class Transition {
    int duration;
    int my_time;
    PlaceList in_places;
    PlaceList out_places;
    ColorSet AllowedColors;

    virtual bool all_marked();
    virtual bool can_fire();
    virtual float firing_probability();
    virtual void tick();
    // two pointers to the scheduling functions:
    bool  (*T_real_time_scheduler)(Trans* t, Color c);
    Color (*C_real_time_scheduler)(Trans* t);
}
```

The method can_fire is used by the executor to check if a transition can fire, and can be redefined to implement different firing rules. Usually its behaviour is not altered because it is general enough to satisfy different needs by changing the implementation of some other methods it uses. In particular, the algorithm on which it is based is the following:

```
bool
Transition::can_fire()
{
    Color color;
    float p;

    // checks for tokens presence and availability
    if (!all_marked()) return False;

    if (trans has more than one firing color)
        color = (*C_real_time_scheduler)(this);
```

```
    if (color is invalid) return False;

    p = firing_probability();
    if (p is 0) return False;  // Trans cannot fire
    if (p is 1) return True;   // Trans can fire

    // There is a conflict: invoke the scheduler on the current trans
    return (*T_real_time_scheduler)(this, color);
}
```

The method all_marked is the ideal candidate to redefinitions: in a coloured transition, this method is redefined to check that the same color is present in every input place. Its default behaviour is to check if a transition can fire, including the check on the Availability Table. The friend function pointed by C_real_time_scheduler manages possible color conflicts appropriately. Its default behaviour is to select the oldest token present in the input places. In this approach the scheduling algorithm is distributed among the transitions, and acts according to local or quasi–local properties of the net, as described further in the paper.

```
float
Transition::firing_probability()
{
    float prob = 1.0;
    int counter = 0;

    for each input place p of trans {
        for each transition in output of p except trans {
            if (transition already scheduled)
                continue
            if (duration of transition is zero)
                continue
            if (transition can fire)
                counter = counter + 1
        }
}
```

```
   if(counter is not zero)
      prob /= counter;
   return prob;
}
```

4.2.1 The basic algorithms for control and simulation

As in many simulation or concurrent languages, the central (virtual) processor is multiplexed among the processes (transitions): this function is performed by a scheduler process which guarantees simultaneous time–flow across all processes by giving a time slice to each in turn. Time is divided in atomic slices which are the only instants in which action may take place, i.e. a transition can either fire or expire. The different management of these time slices results in the development of two separate schedulers: the synchronous and the asynchronous scheduler.

The synchronous scheduler, dedicated to real time requirements, checks each transition in turn at every time instant to see if it is involved in any event, and, in that case, executes the transitions. This approach seems to be particularly suitable in a real environment since it does not require any hypothesis on the duration of each transition. Regularly checking for the completion of a transition activity, it easily copes with the unfixed time delays which may be due to the human factor in a real world application. On the other hand, from the asynchronous point of view, only a set of relevant time slices are examined. These instants are characterized by the following conditions:

1) the initial time;
2) any time slice in which a resource become available according to its availability table;
3) any time slice in which the controller associated to a transition communicates the completion of its task instance, that is the transition has to expire immediately and mark its output places.

At each significant time slice, the firing rule is tested for each transition, and, if verified, the transition is activated and its function, i.e. the code for its controller, executed. In a simulated environment, a transition can remain active for several time slices, according to the user definition of its temporal duration (possibly non deterministic). The controller drives the expiring of its transition activity, sending end–of–activity messages to the transition. A test on the firing rule of all transitions is executed in the time slot immediately following the expiring of a transition, for determining possible transitions which could be enabled by the expiring and marking of transition output places.

The net animator is not dependent on net topology, although particular net constructs caused some complexities on the structure of the algorithm. In particular, the algorithm provides a general–purpose behaviour for conflicts resolution: the firing rule can not be computed locally in a transition, but depends on the state of other conflicting transitions.

The global executor is a loop which is iterated until something happens in the net:

```
for (every transition `t' in the net)
   if (t.can_fire())
      t.fire()

for (every transition `t' in run_queue)
   t.tick()
```

The method fire consumes the input tokens and produces the output tokens for transitions having a null duration. Other transitions are put in the run_queue and will produce the output tokens when the method tick will find that the transition duration has expired.

```
void
Transition::tick()
{
   if (!my_time) return;
   if (!(--my_time)) {
      call task function
      produce output tokens
   }
}
```

The task function can be called whenever appropriate, to deal with the outside world.

4.3 SUBNET HIERARCHIES

Subnets can be defined in tau–nets, and can be used as particular types of transitions. Particular constructs, called Macro Structures (MSs), are defined as a generalization of subnets: a MS is not a topological fixed construct (i.e., a subnet) which can be instantiated assigning variable values, such as token values etc. (as procedures in programming languages), but is a class of topological "similar" constructs, i.e. having some common basic features. MS are defined in terms of high–level frames in the FMS knowledge base. Two instances of the same MS (the Task MS) are shown in figure 4 (Task 1 and Task 2); their common feature is that all cases are conflicts (or alternatives), but their topologies are different.

The set of general MSs available in the system can be easily updated, at the level of FMS knowledge base, introducing new high–level frames, representing some particular behaviour in terms of methods and slots of the frame. Other basic features of the FMS KB are described in a further section.

5. THE SCHEDULING SUBSYSTEM

PN theory does not provide any conflict–solving method: the firing rule does not specify, for example, which of the conflicting transitions must fire and when.

As previously stressed, decision situations play a fundamental role in the net evolution; the decision level should deal with two basic problems:

I)   conflict solving;
II) guiding the net execution towards the (sub)optimum solution.

These two aspects are, of course, deeply interconnected since they both rely on the choice of a path at a decision point, which in most cases is a conflicting situation. Before describing the structure of the scheduler, the conflicting situations are now briefly discussed.

5.1 CONFLICTING SITUATIONS

Some critical situations are possible. They can be classified in the following categories:

I)   a topologic conflict (figure 6a),
II) a color/project conflict (figure 6b),

III) a duration indeterminacy (typically in W–transitions),
IV) mixed situations, typically a simultaneous topologic and color conflict.

Figure 6: Conflicting situations

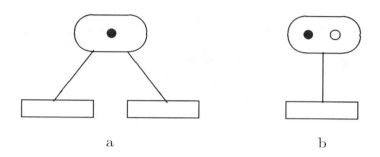

a   b

A topological conflict arises when a single token could enable more than one transition in a mutually exclusive way: a decision must be taken, regarding the transition to activate. In the FMS environment, this means that at least two machines, able to process the incoming pallet, exist; therefore, there is the need to establish a policy according to which the most suitable processing cycle for that particular pallet is decided (see the task1 in the sample example). Note that the decision should also take into account currently running machines (i.e. active transitions in a topological conflict), which will be available and could contribute to a better schedule.

A colour conflict concerns the transitions enabled at a given time, characterized by multiple enabling sets of tokens of different colours. In this case, the decision depends also on token semantics, i.e. it is affected by the role the token plays in the specific problem instance. This means that a single machine could find multiple pallets ready for processing at a given time, and there is the problem of choosing which of them must be chosen first, and which must be delayed. Also in this case, the decision should take into account also possible further tokens (projects) which are not yet available, but will be generated in the ``near future'', by one of the preceding transitions. That is, an enabled project can be delayed for giving precedence to another one which is going to be ready for execution. This allows a sort of ``branch in the future'' in the scheduling process. For example, let us consider figure 4: a project available for execution in place I_queue2, can be delayed to wait for another project currently running on one between the W1 and W2 wait transitions. This can be done very simply using the data structure of the PN, which reflects the topological features of the problem: a similar behaviour is considerably more difficult and inefficient in traditional simulation algorithms, based on simpler data structures, which require time–consuming search processes on tables, for finding the possible connected tasks.

The third critical situation is a typical case of duration with soft constraints, i.e. without a fixed time limit. In this case, since the net is intrinsically deterministic, a decision should be taken on the length of this time duration, in the allowed time span. Soft constraints are used in FMS language to give the decision system a degree of freedom in order to minimize input queues for the machines. Queues may in fact cause stocking problems in real applications, and they should be avoided to limit the costs.

Of course, some of the situations described above may occur simultaneously, and a cooperation mechanism among the strategies should be considered.

A real–time scheduling subsystem has the role of a decision–maker, for the previous cases. It utilizes the knowledge stored in the KB, both on the structure (high–level, definitional component of the KB) and on the features of the particular problem (lower level, assertional component), including the optimizing criteria. The following consideration should be taken into account in the definition of a feasible scheduler: the optimum solution is generally not available in a reasonable amount of time, for real world cases. For example, in our FMS case study, we have to cope with a lot of tasks and resources, even in easy application problems. Therefore, only local (or quasi–local, as we have seen) decision–making processes are practically feasible, to achieve reasonable response times from the system.

5.2 SPECIAL PURPOSE SCHEDULERS

In defining special purpose schedulers, the focus is on flexibility, that is on an easy integration of different scheduling mechanisms. Quasi–local scheduling strategies keeps into account the future time intervals, handling the two ``branch in the future'' cases we have seen for color and topological conflicts.

In the colour conflict case, several pallets could be ready to be processed on the same machine at the same time. We call R this set of tokens. A decision on which pallet should be selected first must be taken.

A naive solution gives precedence to the pallet belonging to R with the most tight schedule. A more sophisticated solution could take into account also possible tokens currently processed by a preceding transition. Let us call F the (possibly empty) set of such tokens. In this case, the selected token could be the one with the most tight schedule in both the sets R and F. If there is more than one token, a precedence is given according to a FIFO policy among those already present in the set R.

The colour scheduler is realized by means of the function C_real_time_scheduler, called by the executor algorithm. This function takes into account the pool of conflicting pallets, and returns the selected one. Two steps are taken to reach the best choice: at first, the pallet with the closest due_date is chosen; then, if no due_date is specified for any pallet, the deadline is taken into account.

This can be expressed by the following algorithm:

```
compute the sets R and F
FirstDue = set of project(s) whose due date is specified and closest
           in both sets R and F
FirstDead = set of project(s) whose deadline is specified and closest
           in both sets R and F
if (FirstDue not empty) {
    select a project prj from FirstDue and present in R
    if (prj does not exists)
        select a project prj from FirstDue and present in F
    return prj
}
if (FirstDead not empty) {
    select a project prj from FirstDead and present in R
    if (prj does not exists)
        select a project prj from FirstDead and present in F
    return prj
}
// the scheduler is not able to choose:
return Failure
```

Another conflicting situation arises when two or more machines compete to get a pallet for processing. In other words, a task could be performed in several different ways and the most profitable combination of human and machine resource must be found. This corresponds a topological conflict. The assumption chosen in this sample scheduler is to fire the fastest transition, unless waiting could result in a potential improvement. This includes the case of waiting for one of the conflicting transitions currently running to complete.

This is not of course the only choice that could be made; other heuristics could be considered, for example to take into account the resources costs, and to combine it with the deadline of the project.
The algorithm for this scheduler is defined in the function T_real_time_scheduler(this, prj).

> detect the conflicting transitions, including those currently running
> select the fastest transition t for color prj
> if (t is already running) return False; // wait for expiring of t
> if (&t != this) {
>   t->fire(); return False;
> } else return True;

Another general problem concerns the definition of compulsory time delays between sequential tasks. Such delays can be generally defined as a time interval indicating the minimum and the maximum delay allowed between the tasks. Since Petri nets superimpose time delays on the projects (there can be conflicts or the desired transition could be already busy), it is often reasonable to adopt the minimum as a fixed time delay.

6. HIGH LEVEL SUBSYSTEM: THE FMS KNOWLEDGE BASE

Temporal and structural properties of FMS problems are represented in a FMS knowledge base, implemented as a structured frame-based multiple inheritance semantic network, described in detail in [10]. At this level, the user manipulates problem entities rather than PN elements. Each problem entity is an object, with related properties and/or methods (procedural definitions). The user interacts with the system by asserting new properties or querying the system about the FMS status.

Therefore, the FMS knowledge base consists of a taxonomic hierarchy of problem entities. Such representation has significant advantages, deriving from their object oriented structure, such as a considerably reduced memory occupation and easy definition of new tasks, projects, or other entities, basing on already existent objects. Besides, the user defines the problem in terms of simple assertions, demanding to the system the task of (automatically) synthesizing the corresponding tau-nets model of the coordination subsystem, as described below. Finally, a taxonomic representation of the problem domain can be a useful platform for the development of the higher level components of the control system of the FMS, the overall scheduling and planning subsystems.

At this high–level, conceptual model, two basic components can be distinguished:
(i) a terminologic component, which models the definitional aspects, i.e. it regards the definition of the general terms and properties of the objects involved in a given problem class (a taxonomy of concepts, including the definition of projects, tasks, optimizing criteria, resource sets, etc.; as an example of the formalism, see figure 7); (ii) an assertional component, regarding particular, individual properties of the objects involved in a given problem instance. A representation for temporal knowledge has been introduced in both components, based on the constructs called tense maps, which allows a partial order on time instants. Time intervals are defined as couple of time instants [11][27]. From a problem definition at this level, it is possible to detect inconsistencies on time constraints between objects in the taxonomy [10]. A detailed description of the Artificial Intelligence methodologies adopted in the design of the knowledge base are described in [10]. In the following we give only a sketch of its structure.

A FMS knowledge base follows the terminologic definitions introduced in the section describing the case study: there are high–level objects, or classes, representing projects, tasks, resource sets, MSs and criteria. Particular projects can be created as instances of the projects class. The projects class contains the basic properties common to any project: a release time (the initial time for a project), a due date (the wished completion time of a project), a deadline (the maximum time after the due date allowed for the project to terminate), the list of tasks involved in the project, the time relations among couple of tasks. Some time relations may be left unspecified by the user; in this case they are supplied by the inference mechanisms of the system (classification, inheritance and special purpose inference algorithms).

The tasks involved in a project are instances of the Tasks class, but inherit also properties from their project: for example, it is possible to introduce a due date and a deadline for each task, but if they are lacking, proper defaults values are computed inheriting from their project. An algorithm has been developed for determining proper defaults values, starting from the project values and from the time relations among tasks.

Figure 7: A fragment of the FMS knowledge base for the case study

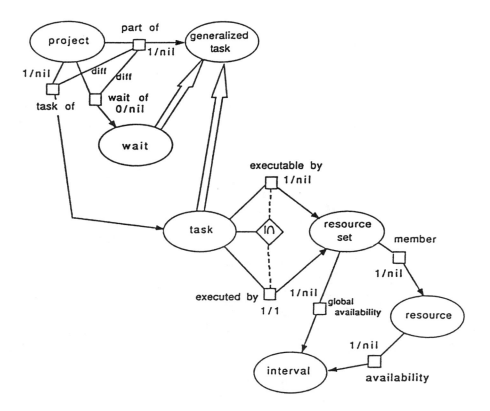

In figure 7 the structure of a fragment of the FMS knowledge base for the problem of figure 2 is shown.

6.1 AUTOMATIC SYNTHESIS OF THE TAU-NETS MODEL

Usually, the task of constructing the PN model of the coordination subsystem is left to the designer; in some cases, powerful graphic aids are available in the implemented software systems [20]. Our approach allows also the automatic synthesis of a tau-model of the FMS, starting from the FMS knowledge base. Besides the objects/frames representing the projects and related knowledge, the system embeds the definition of different frames for the representation of MSs: the main are the Task MS and the Wait/Connection MS. They are sub-classes of the MSs class. A Task MS embeds the definition of its TIs, its related resource sets and its I/O queues.

A Task MS behaves as a macro–place, a Wait/Connection MS as a transition. The tau–nets model of the example is shown in figure 4. It is composed by a sequence of a Task MS instance and an instance of Wait/Connection MS, followed by another instance of the Task MS. Wait/Connections MS are automatically created starting from the temporal relations among the tasks of a given project. Task MSs are created starting from the available tasks and resource sets properties represented in the FMS knowledge base. During this construction process, as many TI–transitions as the number of resource sets are created for each task; each transition has a set of input places corresponding to its resource sets. Each set of the TIs for a given task is successively connected both to an input queue (a place connected in input to all TI–transitions of the given place), and to an output queue (a place connected in output to all TI–transitions of the given task). A proper algorithm for the creation of a tau–nets model for the coordination subsystem explores the FMS knowledge base to build the MS instances and their connections as previously explained.

The result of the execution of such an algorithm given as input the initial problem instance automatically produces the tau–nets in figure 4. A further graphic algorithm automatically defines the Cartesian coordinates for each node in the tau–nets.

The initial marking of such tau–nets is determined by the following algorithm:

– one neutral token is assigned to all the places available to all projects;

– resource places that can be used only by a limited set of projects will have one token for each color/project, according to their definition in the problem instance Knowledge Base;

– for each project, it will be assigned one token (or more, depending on the number of workpieces of the same type to be processed) with the proper color to the input queue(s) of its initial task(s) (in the example, colours P1 and P2 in the place I–Queue1).

7. CONCLUSION

We presented a class of algorithms for the sub–optimal solution of a particular class of problems of process scheduling. The global hybrid approach for the problem specification has been discussed. A case study, regarding the design of the coordination and the real–time scheduling subsystems of a FMS, has been adopted in the paper. The approach described in this paper is based on extension of PNs to embed an object–oriented paradigm, and on structured knowledge representation paradigms. The paper focuses on the former, i.e. tau–nets model for process scheduling.

Tau–nets have been introduced as a modelling tool for the coordination subsystem, defining its major features. The method for the automatic construction of a tau–nets model starting from a FMS knowledge base has been described.

Basic algorithms and ideas have been presented for the dynamic behaviour of the coordination subsystem and for the scheduling policies of the real–time scheduling subsystem. In particular, the algorithms for an original asynchronous discrete events simulator/animator of tau–nets has been described in detail.

The architecture described in this paper has been implemented in different versions, running on both Unix and PC/MS–DOS platforms [10][28]. The software system, implemented in C++, is derived from the PETREX system [29] and from FMS–PTX [24]. The Unix version is currently used in a joint industrial project for the simulation and real time control of a large–size system (a blast furnace). This version is implemented on a DEC workstation 5000/240, under a real–time asset of Unix (DEC RT–Ultrix).

ACKNOWLEDGEMENTS

We acknowledge many useful discussions with Marcello Frixione, Francesca Gandolfo, Cesare Mastroianni, and Renato Zaccaria. This work is partially supported by Special Projects on Robotics and Artificial Intelligence of the Italian Ministry of University and Research (MURST 40%) and of the National Council of Research (CNR–PFI2, grant no. 18/89.00562.67/115.14231).

REFERENCES

[1] J. M. Anthonisse, K. M. van Hee, and J. K. Lenstra. Resource-constrained project scheduling: an international exercise in dss development. Note OS–N8702, 1987.

[2] M. J. Gonzales Jr. Deterministic processor scheduling. Computing Surveys, 9, Sept. 1977.

[3] E. L. Lawler, J. K. Lenstra, and A. H. G. Rinnooy Kan. Recent developments in deterministic sequencing and scheduling: A survey. In Proc. NATO Advanced Study and Research Institute on Theoretical Approaches to scheduling Problems}, Durham, England, July 1981.

[4] S. R. Faulk and D. L. Parnas. On synchronization in hard–real–time systems. Communications of the ACM, 31(3):274--287, March 1988.

[5] A. K. Mok. The design of real–time programming systems based on process models. In Proc. IEEE Real Time Systems Symposium, pages 5–17, Austin, Texas, Dec 1984. IEEE Press.

[6] G. Loy. Designing an operating environment for a real time performance processing system. In Proc. 1985 Int. Computer Music Conf., pages 9--13, Burnaby B.C., 1985.

[7] R. B. Dannenberg. A real time scheduler dispatcher. In Proc. 1988 Intl. Computer Music Conference, pages 239--242, Cologne, Sept.19–24 1988.

[8] D. P. Anderson and R. Kuivila. A system for computer music performance. ACM Trans. on Comput. Sys., 8(1):56--82, 1990.

[9] J. Xu and D. L. Parnas. Scheduling processes with releases times, deadlines, precedence, and exclusion relations. IEEE Trans. on Software Engineering, 16(3):360--369, 1990.

[10] A. Camurri and M. Frixione. Structured representation of FMS integrating SI–nets and high–level Petri nets. Applied Artificial Intelligence – An International Journal, 4(2):109--131, 1990.

[11] J. F. Allen. Towards a general theory of action and time. Artificial Intelligence}, 23:123--154, 1984.

[12] A. Pagnoni. Stochastic nets and performance evaluation. In W. Brauer, W. Reisig, and G. Rozenberg, editors, Lecture Notes in Computer

Science, Advances in Petri Nets 1986, volume 254. Springer–Verlag, 1987.

[13] J. L. Peterson. Petri Net Theory and the Modeling of Systems. Prentice Hall Inc., New York, 1981.

[14] C. A. Petri. General net theory. In B.Shaw, editor, Proceedings of the Joint IBM and University of Newcastle upon Tyne Seminar, pages 131–– 169, 1976.

[15] K. Jensen and G. Rozenberg, editors. High–Level Petri Nets. Springer Verlag, 1991.

[16] F. Feldbrugge and K. Jensen. Computer tools for High–Level Petri nets. In K. Jensen and G. Rozenberg, editors, High–Level Petri Nets, pages 691––717. Springer Verlag, 1991.

[17] M. Baldassari and G. Bruno. Protob: An object oriented methodology for developing discrete event dynamic systems. In K. Jensen and G. Rozenberg, editors, High–Level Petri Nets – Theory and Applications, chapter 25, pages 624––648. Springer Verlag, Berlin, Heidelberg, 1991.

[18] V. Pinci and R.M. Shapiro. An integrated software development methodology based on hierarchical colored Petri nets. In Proc. 11th Intl. Conf. on Application and Theory of Petri Nets, Paris, 1990.

[19] A. Sahraoui, H. Atabakhche, M. Courvoisier, and R. Valette. Joining Petri nets and knowledge based systems for monitoring purposes. In Proc. IEEE Intl. Conf. on Robotics and Automation, pages 1160––1165, Raleigh, North Carolina, March 1987. IEEE Computer Society Press.

[20] G. Balbo, G. Chiola, G. Franceschini, and G. Molinar Roet. Generalized stochastic Petri nets for the performance evaluation of FMS. In Proc. IEEE Intl. Conf. on Robotics and Automation, pages 1013––1018, Raleigh, North Carolina, March 1987. IEEE Computer Society Press.

[21] J. Martinez, P. Muro, and M. Silva. Modeling validation and software implementation of production systems using high–level Petri nets. In Proc. IEEE Intl. Conf. on Robotics and Automation, pages 1180––1185, Raleigh, North Carolina, March 1987. IEEE Computer Society Press.

[22] J. C. Gentina and D. Corbeel. Coloured adaptive structured Petri–net: A tool for the automatic synthesis of hierarchical control of flexible manufactoring systems (F.M.S.). In Proc. IEEE Intl. Conf. on Robotics and Automation, pages 1166--1173, Raleigh, North Carolina, March 1987. IEEE Computer Society Press.

[23] A. Camurri, A. Poggi, G. Vercelli, and R. Zaccaria. A–nets: Structured representation of time and actions using Petri nets. In European Workshop on Application and Theory of Petri Nets, Venice, Italy, June 1988.

[24] A. Camurri, P. Franchi, and F. Gandolfo. A timed colored Petri nets approach to process scheduling. In Proceedings of IEEE COMPEURO–91 Intl. Conference, pages 304--309, Bologna, Italy, May 1991.

[25] M. Ajmone Marsan, G. Balbo, and G. Conte. A class of generalized stochastic Petri nets for the performance evaluation of multi–processors systems. ACM Trans. on Computer Systems, 2(2):93--122, May 1984.

[26] K. Jensen. Coloured Petri nets and the invariant method. Theoretical Computer Science, 14:317--336, 1981.

[27] Y. Shoham. Temporal logics in AI: Semantical and ontological considerations. Artificial Intelligence, 33:89--104, 1987.

[28] A. Camurri and P. Franchi. An approach to the design and implementation of the hierarchical control system of FMS combining a structured knowledge representation formalism and high–level Petri nets. In Proc. IEEE Intl. Conf. on Robotics and Automation, pages 520--525, Cincinnati, Ohio, May 1990. IEEE Computer Society Press.

[29] A. Camurri and E. Troiano. Petrex: a software tool for system modelling and simulation based on Petri nets. In Proc. AICA Annual Conference, Trieste, Italy, 1989.

SECTION III:

SCANDINAVIAN MANUFACTURING AUTOMATION IN REAL APPLICATIONS

Manufacturing Automation at the Crossroads
L.F. Pau and J.-O. Willums (Eds.)
IOS Press, 1993

Chapter 7

Implementing Manufacturing Platforms in Scandinavian Industry

Jan Storve and Stig Ulfsby
Avenir A/S
Oslo, Norway

Abstract: This chapter evaluates the usefulnes of manufacturing platforms as seen from a Scandinavian industry perspective. It highlights the requirements that has to be met for such platforms to be cost–effective tools in manufacturing.

It finally compares the potential for such platforms in the different Scandinavian countries.

1. WHAT IS A MANUFACTURING PLATFORM?

A manufacturing platform is a software platform for developing software for manufacturing automation. That is a set of software modules and interfaces which can be used in different projects with different hardware and software environments.

The main idea is that the application–processes work against a standardised interface, and thereby are independent of system specific environments. The application–process can be a cell–controller, a data capture program etc. System specific environments are:

- Hardware.
- Operating system.
- Database management system.
- Data communication.
- Devices like manufacturing machines and data capture equipment.

The application programmers interface to the rest of the system is, as indicated independent of the environments. Application programs can therefor easily be moved to other environments assumed that the manufacturing platform is available in these environments.

In addition application processes that communicate via the message handling interface, in principle can be distributed on several CPUs, in one computer or different networked computers. The application program will be independent of the number of CPUs (except for response times). A manufacturing platform can include:

- Real–time services like message handling, shared memory, access control and event handling.
- User interface tools.
- Development test–tools.
- System administration tools.
- Communication drivers for devices.
- Application processes like event logging and alarm system.

A manufacturing platform can be considered as an operating system for manufacturing automation software. Mainstream, which is described in another chapter, is an example of a manufacturing platform.

2. THE POTENTIAL OF A MANUFACTURING PLATFORM.

The idea of a manufacturing platform is in accordance with the trend toward standardisation and more open systems. If the platform supports several environments and hardware platforms, the following are the potentials:

2.1. PORTABILITY.

Application programs using a manufacturing platform, will, because of the environment independent interfaces, be portable. In principle the application can be moved from one environment (computer, operating system, DBMS, data communication) to another without any changes. It is not necessary to make special solutions for different environments.

2.2. VENDOR INDEPENDENCE.

Since the application programs are portable, one becomes more vendor independent. It will be less expensive to change vendor because the applications do not have to be rewritten. One will however be dependant on the vendor of the manufacturing platform. In a far developed manufacturing platform it will also be possible to have applications that are distributed on networked hardware from different vendors.

2.3. REPEATABILITY.

Portability will lead to repeatability. That means developed application modules can be reused in other applications. In other words, it is not necessary to do the whole developing work again due to new hardware or operating system etc.

2.4. DISTRIBUTE ABILITY AND FLEXIBILITY.

If the manufacturing platform allows the application processes to communicate through a general message exchange mechanism, then the processes can be distributed on different computers in a network. The application process can be moved to another computer at a later stage to trim the system or fit it to changed conditions. New application processes can easily be included through the systems lifetime.

2.5. MODULARITY.

With a common platform, the application processes can be developed by several co–operating software houses. Common interfaces makes "plug–in" software–modules possible.

2.6. COSTS.

Repeatability, portability and modularity makes the system development process more efficient, and will reduce the development costs.
On the other side, a manufacturing platform costs, and it costs to learn it and to maintain it. For small projects it is therefor not cost efficient to use one. Even if the application is small, the whole system becomes large due to the size and complexity of a manufacturing platform. How large the project must be to make the use of a manufacturing platform cost efficient is dependant on its price and maintenance costs and the productivity gained.

2.7. RISK

The use of a manufacturing platform will make it easier to decide if an application can be developed within a frame of cost and time in a given environment, and with a reasonable degree of certainty know that the results will be working according to the specification. In other words the uncertainty will decrease.

2.8. QUALITY.

Application programs will be smaller using a platform. There will be less code to test, and the platform will have test tools making it easier to test. If one application module can be used relatively unchanged in several applications, the testing can be done once for all. A new project will take advantage of a module that is already tested in production in another project. The probability of errors will be smaller than without a platform.

3. THE ROLE OF A MANUFACTURING PLATFORM IN SCANDINAVIAN INDUSTRY

A good manufacturing platform is a tool for efficient and standardised development of high quality real–time software for manufacturing automation. The complexity and cost of such a platform sets however some requirements to make it profitable to use:

- The projects must be large enough to justify the increased costs and complexity.
- The number of projects must be large enough to gain from reusability and the cost of building up know–how on the platform.
- The manufacturing platform must cover a great variety of applications.
- A representative number of application developers must use the same platform to gain from portability and get a milieu around it.

Very few Scandinavian manufacturing companies fulfil these requirements. The users of a manufacturing platform should therefor be consultant companies to have enough projects that can gain from a platform.

Even for consultant companies the market is small. In Norway we experience that there are very few manufacturing companies, that are large enough to have manufacturing automation projects that can gain from a manufacturing platform. Most large Norwegian companies are process industry or offshore industry that have other kind of automation.

In Denmark the market is some better, but still small. In Sweden however the market is larger with a reasonable number of large manufacturing companies that can gain from a manufacturing platform. A consultant company that has Scandinavia as market should therefor have a large enough market to gain from a manufacturing platform. Also the largest manufacturing companies like Volvo and SAAB can gain from it by internal use.

Chapter 8

Computers in Manufacturing at Danfoss A/S

Jørgen Nimb
Danfoss A/S
Grønvejen 2
DK–6340 Norborg, Denmark

Abstract: The paper summarises the experiences of introducing data–based manufacturing concepts in Danfoss, one of the INTEGRA members and the largest manufacturing group in Denmark. It outlines the trend in planning systems towards buying standard software both on main frame computers and on distributed PC systems, and more dedicated systems like computer aided warehouse systems via so–called open systems with local relational databases and PCs connected to LAN's. The present focus at Danfoss is the implementation of Japanese manufacturing principles, in order to improve the process capabilities, introduce excellent maintenance of all equipment, achieve quick changeover, and eliminate all kinds of waste in the manufacturing plants. The paper underlines that it is "the people on the manufacturing floor who make everything happen", not "the systems". The most important basis for success in manufacturing is involvement, participation and motivation of the people in the plants.

1. INTRODUCTION

Danfoss is a leading producer of precision mechanical and electronic components and of intelligent "mechatronic" devices. The company has modern factories on four continents, subsidiaries and representatives in more than 100 countries, and 14,000 employees worldwide.

Danfoss has production facilities around the world capable of turning out some 80,000 components a day. All are geared for efficient production runs and make use of the most advanced manufacturing techniques.

Moreover, all aim for total quality control. Dedicated quality assurance teams – independent of scheduling and production requirements – ensure strict adherence to customer specifications. The teams participate in development work, monitor every step of production and conduct demanding tests of finished goods.

Naturally, the company also employs the very latest in quality control equipment. Danfoss was one of the first in its field to obtain ISO 9000 certification – the quality control certificate of the International Standards Organization.

The company's strengths in production and quality control are matched by equally outstanding capabilities in research and development. Danfoss has pioneered techniques like load–sensing hydraulics, intelligent refrigeration controls, radiator thermostats, CFC–free compressors, and thermostats for household appliances.

As a result, Danfoss today defines the state of the art in many areas of components technology.

2. THE DEVELOPMENT UNTIL NOW

At Danfoss we started talking seriously about computer integrated manufacturing (cim) in the middle of the eighties. Even though we had used that kind of technology for some time to a fairly high extent, we had not used the phrase broadly in our production environment.

We used CAE for Finite Element Analysis, simulation and software development, as well as CAD, CAW (warehousing), CAR (repair). And we extensively used an extremely efficient in–house built Manufacturing Resource Planning system (MRP). Most of our systems were – and are – host based (IBM). Investments were made in decentralised systems for electronic lay–out of printed circuit boards and Application Specific Integrated Circuits (ASIC).

In the shop floor of the factory we also invested heavily in CNC–machines of all kinds but none of these single machines were tied together in local area networks (LAN). We still used paper tape as a basis for our production. At that time there was much talk in the engineering community that it was 'a must' to implement total integration all over the shop floor. At Danfoss we listened reluctantly – and we are glad we did. It meant that all the time we had been able to produce. It meant that we were in a position which made it possible for us to focus our interest and our resources on the important points in the shop floor.

Still, at that time we also started to make some preliminary work with robotics in mechanical assembly, and we gained our first experience with a new assembly technology for electronics, called surface mounted technology (SMD).There was also a great fuss about flexible machining cells and systems (FMC & FMS). But altogether we tried to make best use of the well known and safe technology. The question which we kept asking ourselves was: "Why should we invest so much money in integrating everything? What can we expect to earn from that way of doing things?" We never tried to solve the computermaker's problem: His sales! All the time we tried to solve our own problems as regards functionality, quality and the dawning interest in our environment. For example a surface treatment that is more neutral to the environment than the one used so far.

3. STATUS

During the last five to seven years the technology development has proceeded at a still higher rate. New computer systems are emerging. They run faster than ever at an almost imaginary low price. They have become commodities to our engineering society – and to the rest of our society. But the software systems, i.e. new concepts of controlling production, new concepts of interconnecting different software packages etc. cannot keep pace with the computer development. The software systems are at the very beginning of a useful openness and ability of true integration.

To us at Danfoss this means that we must be very sensible and again ask the question: "Why should we allocate all these resources to computer integration?" Anyway, we gradually install LAN's in shop floors where we can calculate and/or 'feel' the positive turnover. We are trying to take a pragmatic attitude towards all PR and campaigns from the big computer manufacturers. Mostly, things must be proven to be implemented at our factories, and we like to do things one step at a time. Especially in the somewhat chaotic marketplace as we see it these last couple of years.

The state of our production environment is fast changing, though. We are amalgamating different production facilities to joint ones. We are also separating and selling facilities in order to match our business system with the actual demand from the market. In our physical lay-out of the production lines we must therefore increasingly stress the terms: Flow orientation, simplicity, and flexibility. In turn follows that the need for computer power will rise less drastically –everything else set equal.

The trend in planning systems is exclusively towards buying standard software both on main frame computers and on distributed PC systems. More dedicated systems like computer aided warehouse systems are now implemented via so–called open systems (which they are not) with local relational databases and PCs connected to LAN's.

The trend in CNC machining is towards more implementation of computer power in the machining centers themselves. Again this means a greater ability for the single machining cell to act in a flexible way without the need for outside control. Even CNC programming can be done directly on the machine while it is producing other parts, and the programming can be done by the operator who knows the machine best. Furthermore the operator will be able to simulate his CNC–programme on the attached CAD–screen before putting the code into work.

4. THE HUMAN SIDE

Investment in information technology is one way of getting higher productivity and achieving competitiveness, but it is not necessarily the most efficient way.

At the moment the most important development in the manufacturing plants in the Danfoss Group is the implementation of the well–known Japanese manufacturing principles.

The objective of these improvement activities is to manufacture the products at the highest possible quality level at the lowest possible costs. This is done by improving the process capabilities, introducing excellent maintenance of all equipment, achieving quick changeover, eliminating the possibilities of doing things wrong, and in general eliminating all kinds of waste in the manufacturing plants.

Dealing with improvement in production it is important to remember that it is "the people on the manufacturing floor who make everything happen", not "the systems" and especially not the computer systems. Therefore the most important and fundamental basis for success in manufacturing is involvement, participation and motivation of the people in our plants.

In order to achieve this a lot of work must be put into communication of the goals, discussions on the problems which people have, systematic teamwork on finding the causes of the problems and set up ideas for solutions, and at last implement and follow up.

Some of these ideas and implement solutions can be computer based, if people in the plants find it suitable, but all in all this has a minor bearing on being a highly competitive manufacturer.

Chapter 9

The possibilities and pitfalls of CIM in medium sized Norwegian Enterprises.

Stig Ulfsby,
Avenir A/S,
P O Box 6824 St. Olavs Plass
N 0164 Oslo, Norway

Abstract: Avenir, as a founding member of the INTEGRA project, has been an active partner with Norwegian industry for over 30 years in introducing advanced technology and modern manufacturing methods. This paper reviews the experiences gained by helping to introduce manufacturing automation in medium–sized Norwegian Enterprises, and in developing a Norwegian strategy for CIM. A survey of medium sized companies showed that even small and medium sized companies have a need for CIM even though in smaller scale than the large companies., and concludes that standardisation is necessary to expand the acceptance of CIM solutions.

1. CIM – THE MOST IMPORTANT TECHNOLOGY FOR INDUSTRY

Computer Integrated Manufacturing is in USA, Japan and the large European industry–nations regarded as the technology that will have the greatest influence on the industry in the near future. The increasing competition and requirement to the products increase the need for using advanced technology. CIM is the answer to many of the challenges the industry meets in this situation.

CIM is a technology that has the purpose to increase productivity, flexibility and quality and to reduce delivery times and bound capital. CIM is a concept that means co–ordination and integration of the company's total applications of information technology and automation.

1.1. NORWEGIAN INDUSTRY LOOSES COMPETITIVENESS.

The industry in Norway that is running best is the primary industry, like oil–, aluminium–, and wood processing–industry. This is due to the natural advantages as oil in the North Sea, cheap hydroelectric power and large woods. It is however in the manufacturing industry the large increase of value is, but the Norwegian manufacturing industry has become less competitive. This is due to the lack of compensation for the high level of salaries by advanced technology.

Norway is one of the countries with the highest labour costs in the world. The competition from abroad is becoming harder on a more international market. EC's open market is increasing the international competition. If Norway is going to keep the high level of salaries and at the same time increase the competitiveness, Norway must be at the edge of applying technology. It must be developed products that can not be developed or manufactured manually. Automation must be done to compensate for the high level of labour costs. Norway can not compete on the premises of the low cost countries, if so, the competition is lost from the beginning.

The market is putting higher and higher demands to shorter delivery times, higher quality and more product variants. One is not satisfied with a grey standard telephone. People want different models and colours. The products life time is decreasing, implying new products to be developed and more often change of production.

1.2. CIM INCREASES THE COMPETITIVENESS.

CIM is reckoned to be the most important technology for the manufacturing industry. This is stated by the Roundtable for European Industrialists and by Stanford Research Institute – International on the authorities of Royal Norwegian Council for Scientific and Industrial Research (NTNF).

The US National Research Council has done an investigation among companies in USA (3). Experiences from these companies shows the following effect on CIM technologies:

- 15–30% reduction of design costs.
- 30–60% reduction of total lead time.
- 30–60% reduction of goods in process.
- 40–70% increase of productivity.
- 2–3 times increase of productive operational time.
- 2–5 times the increase of quality.

Messersmidt Boelkow Blohm in Germany has integrated 25 numerically controlled machining centres, automatic tool transport and tool change in a flexible manufacturing system (FMS). Experiences from this shows that FMS compared to separate NC machines gives the following results:

- The number of tools reduced by 52%.
- Labour reduced by 50%.
- Machining costs reduced by 30%.
- Lead time reduced by 20%.
- Invested amount 10% less than for separate NC machines.
- Yearly costs reduced by 24%.

Similar experiences from a FMS at General Electric in USA shows:

- 240% increase of productivity.
- 30% reduction of area.
- Reduction of lead time from 16 days to 16 hours.

These numbers, taken from large international companies, show that CIM can be profitable. We will later look at some examples from Norwegian companies.

2. POSSIBILITIES OF CIM.

CIM is a concept that implies integration of information systems and automation in manufacturing, from idea to completed product. CIM implies the co−ordination of the company's total applications of information technology an automation.

CIM is often used as a common term for the technologies that involves use of information technology in manufacturing. E.g. CAD/CAM, materials and production planning, manufacturing automation, data capture, store automation and administrative information systems.

The new with CIM is the integration of automated islands in the company. The usual situation today is that each department has its own systems. The design department has a CAD system, the planning department a MRP system and the production department a CAM system. All the systems work as automated islands. CIM implies that the islands are integrated to a whole.

With integration we mean that information from one system shall be available for other systems without manual transfer. E. g. the parts list from the CAD system must be available for the materials and production planning system, and the production plan must be available for the shop floor control system and the flexible manufacturing system.

In addition to the effect of every single system, integration will give a set of new advantages:

- The information flow in the company is made more effective. The manual information handling is strongly reduced. Information is available faster and the costs of information handling is reduced.

- Errors in information is reduced. Manual transfer of information is an error source. Such errors may not be discovered until the customer has received the order, which causes severe costs and loss of confidence.

- Better decision support. Better availability of information from several systems in the company makes the basis for decisions better.

- Increased automation. The co–ordination of several production automates, transport automates and store automates increases the total degree of automation. Further it opens for automation of operations which earlier was not appropriate to automate.

- Set–up times are reduced. Instead of manually changing the programs of the production automates, new programs can be loaded automatically from a superior controller.

- Reduction of total lead times. The higher efficiency of information flow, the increased automation and the reduced set–up times all reduces the total lead time.

3. NORWEGIAN INDUSTRY'S NEED FOR CIM.

In connection with the development of a Norwegian strategy for CIM, a survey of the demands for CIM in Norwegian industry was made. Medium sized companies was the major group in the survey. A clear conclusion from this survey is that most companies has demands for CIM. Even small and medium sized companies has a need for CIM even though in smaller scale than the large companies. The following is a summary of the survey.

3.1. INTEGRATION IS MORE IMPORTANT THAN BETTER SUBSYSTEMS.

Today systems are used for several applications: CAD/CAM, materials and production planning, simulations, factory automation etc. It is still much to be done on these areas, but a clear conclusion from the survey is that the development of such systems has come so far that it is more important to make them work together than to improve the single systems.

It must be possible to use information from one system further in other systems without manual transfer of data. This will increase the efficiency of the information flow in the company and reduce the possibilities for errors, and make it possible to utilise the systems more efficiently.

3.2. STANDARDISING NECESSARY.

There is no vendor that can deliver all the systems a company needs. In a CIM solution it is therefore necessary to make systems from different vendors to work together, even on computers of different brands. In the same way production machinery and robots from different vendors must work together.

This requires standardisation of interfaces between systems and between machines. It requires standards for data communication and neutral formats for transfer of information.
TOP (Technical Office Protocol) and MAP (Manufacturing Automation Protocol) are examples of standards for data communication which is developed for information systems and production equipment respectively. These standards defines how data is to be transferred between different computers and production equipment on a local area network, but they do not define which kind of information is to be transferred.

Examples on standards for structuring information to be transferred is IGES, PDES and EDIFACT. IGES (Initial Graphics Exchange Specification) is a standard for exchange of drawings and geometry between CAD systems. PDES (Product Data Exchange Specification) is a standard for transferring more information about the product than graphics and geometry. EDIFACT is a standard for exchange of commercial information like purchase order and invoice between systems in different companies. Except for these standards, there are few standards for exchange of information between different systems.

3.3. ORGANISATION IS A LARGER PROBLEM THAN TECHNOLOGY.

The companies do not feel that lack of technical solutions is the main problem for introduction of CIM, but rather the organisation.

The company lacks a strategy for CIM. The design manager buys a CAD system, the production manager buys a CAM system and the planning manager buys a MRP system but no one has the responsibility to ensure that they can work together.

The management does not understand the consequences of CIM and the possibilities of CIM. It may be difficult to calculate the economical advantages of CIM. Often new technology is met by sceptics among the older employees. They feel new technology as a threat against their position because they don't understand it.

3.4. CIM MUST BE A PART OF THE COMPANY'S STRATEGY.

CIM will have great influence on most of the organisation. It is therefor impossible to realise CIM without making it a part of the company's strategy. The management must have a conscious stand to CIM and utilise the technology to gain competitive advantages.

3.5. CIM IS IMPORTANT IN ALL TRADES.

CIM is equally important in all industries. There are no great differences in needs between companies in different trades as mechanical industry, electronic industry or furniture industry. The systems will be different from trade to trade, but the needs for utilising the technology, co–ordination of systems and integration are the same.

4. EXAMPLES FROM MEDIUM SIZED ENTERPRISES.

4.1. CASE 1 : NOREMA A/S

STOCK IS REDUCED TO HALF BY COMPUTER CONTROLLED ORDER PRODUCTION.

Norema is making fitting out for kitchens, bathrooms and wardrobes. 450 sections are made a year. There is 350 employees at the factory and 100 in sales. The sale is solely to the Norwegian market, and the company has a market shear on about 1/3 of the Norwegian market. Due to the situation in Norwegian building industry, Norema now goes for export.

Norema has a modern factory with advanced equipment which is not utilised well enough. Through a CIM project Norema wanted a better co-operation between people and integration of systems, machines and routines. The project was focused on:

- Increasing the delivery service.
- Reducing capital bound in materials and manufactured goods.
- Reducing fault costs by reducing wreck and discovering errors earlier.
- Better capacity by reducing spill time for bottleneck machines/ operators.
- Better overview of status in production.
- Better motivation of employees.

Order production has been used for longer time at Norema for assembly of sections. The production of parts was however done in series to store. By an automatic production planning system, the part production is now also order based. This has reduced the store of parts to less than the half.

Order based delivery of elements from a supplier is also introduced. This factory produces now complete spray painted fronts directly to the assembly line at Norema. The store of fronts are thereby reduced to one third. The delivery is one according to a plan generated by the computer system.

Order based packing has reduced the stock of manufactured goods with 60%, and the number of shortages with 80%. The demands for packing is daily defined for each packing machine from parameters like stock level, demands for the nearest seven days and the safety stock. Earlier the packers got a packing plan for a week. Now they every morning get a list from the computer telling what to be packed during the day to avoid shortage and forwarding. The employees know they are packing the right things to the right time.

Norema is installing a completely automatic assembly line. This will be connected to the production planning system and assemble the sections based on customer orders.

4.2. CASE 2: ABB NATIONAL TRANSFORMER, SARPSBORG

FULLY INTEGRATED ORDER HANDLING.

ABB National Transformer, Sarpsborg has 180 employees and manufactures industry transformers. These are complex and usually customer specified. They make 60–80 transformers a year, and seldom more than three equal. They therefore make 30–40 new designs a year.

By use of IT, fully integrated order handling in realised in the company. A common database is created in order to increase the efficiency of the routines for sales, calculations, design, manufacturing, purchase and economy.

The delivery time is reduced from 30 to 20 weeks and the lead–time in manufacturing from 50 to 40 days.

The company focuses on a "best service strategy". The customers shall have a fast response with a high degree of flexibility – with focus on custom design. One of the main results of the CIM project is a system for fast development of offers.

Earlier several departments was involved i developing an offer. The sales department handled the customer, the design department made a design, the manufacturing departments estimated the manufacturing costs and the materials department estimated the materials costs. It could take weeks to make an offer.

Now the offer is made nearly automatically by a computer and an offer can be made within a few hours. On basis of some main figures about the transformer, a CAD system makes a design, and work hours and bill of materials are computed based on the design. The different departments are responsible only for the maintenance of the costs of man–hours and materials and a parts library in a common database. Now only the sales and calculation departments are involved in the offering.

The common database has also made a lot of information available for most of the employees. E.g. the service department put in information on everything that is done with a transformer after it is delivered, and all test data is put into it. The sales engineers has a portable PC brought with them when they are on customer visit, so the customer can get answers at once by connecting on–line to the machine in Sarpsborg. Also the vendors can be connected to the CAD system to take out drawings. In this way the company saves a lot of time which is often a critical factor.

4.3. CASE 3 : ABB DATAKABEL.

ABB Datakabel A/S is a subsidiary of ABB Norsk Kabel A/S. They are making data cables (coax, telephone cable etc.) and have about 180 employees. They have about 4000 variants of which 400 is on stock. 20% of the production volume is customised products, the rest is production for stock.

ABB Datakabel has made an integrated solution of order handling, design, production scheduling and data capture. The aim is to reduce lead–times and increase production capacity and customer service. A central system in their CIM concept is production scheduling.

The sales department has an order system used to follow up the customers and register the orders. This is integrated with the material management system which keeps track on what is on stock. When stock level is below security level, a production order is set out. This order is transferred to the scheduling system. The scheduling system takes the product data from the product description system, brakes the order into machine activities and makes a production schedule for all the orders.

A data capture system gets information from the factory floor how much is made on each operation of each order and how much time it has taken. This information is used by the scheduling system to react on the actual situation on the factory floor.

5. PITFALLS.

Lack of strategy is a main pitfall to fall in when realising CIM. Not only a technology strategy is enough. A logistics strategy and production strategy is needed. The main rule simplify – automate – integrate should be followed. To just automate manual routines and the traditional way of doing things does not work.

Simplify first! Is it necessary to make all the variants the company is manufacturing? Find out which are the most profitable. Standardise parts and raw materials! This will reduce the stock and simplify the materials flow. Simplify the routines! Using a common database may simplify the information flow and the routines.

After simplifying the materials flow and the routines, automation can be done. Finally, to get the full utilisation, integration can be done.

Another pitfall is to focus too much on technology. The technology may be so interesting that the technology itself becomes the goal, and not means to reach the company's business goals. Put the company's business goals in focus, and make the CIM strategy from that point of view.

SECTION IV:

GUIDELINES FOR INTRODUCING MANUFACTURING SYSTEMS :

THE INTEGRA PROJECT RECOMMENDATIONS

Manufacturing Automation at the Crossroads
L.F. Pau and J.-O. Willums (Eds.)
IOS Press, 1993

Chapter 10

GUIDELINES FOR SELECTION OF MANUFACTURING SOFTWARE

Ricardo Tosini
MSD Spa
Torino, Italy
Jan–Olaf Willums
Nordic Manufacturing Forum
Oslo, Norway

Abstract: One purpose of the INTEGRA project is to make a set of guidelines for the evaluation and selection of manufacturing software available to planners, implementers and users of such software. The guidelines summarize the experiences made by EUREKA Project INTEGRA, adapted to the specific needs of the other INTEGRA partners. Italian partners in the paper addresses both technical professionals and business executives.

1. THE NEED FOR GUIDELINES IN SELECTING MANUFACTURING SOFTWARE

The promise of Computer Integrated Manufacturing (CIM) is based on the premise that the whole is greater than the sum of the parts. In simple terms, from an integration viewpoint, the enterprise consists of three parts:

- business functions
- product engineering
- manufacturing

Automation has had a profound impact on the business and engineering components of the enterprise. The business segment was quick to recognize and embrace automation's potential. But the product engineering segment has also embraced automation in the form of CAD/CAE systems, with acceptance being spurred by the rapid price/performance changes of technical workstations.

Manufacturing, the most complex of the three components, is now the focus of automation efforts. Early efforts looked at the return–on–investment substitution of capital investment for direct labour and material costs. This led to CAM (Computer Aided Manufacturing) systems being adopted on an ad–hoc basis, and in many cases these decisions were insular and highly decentralized in their perspective.

It is now clear that greater benefits accrue from a manufacturing automation strategy that not only integrates automated operations within the manufacturing domain but also integrates the manufacturing domain with the engineering and business domains, which are further down the automation path. Thus, managers of automation projects should look not only at the capability of software to meet the immediate needs of projects, but also to the ability of software to support the trend toward integration within the enterprise. This is also a logical step towards the fully integrated enterprise, a concept that foresees a much stronger linkage between the technical, administrative and general management aspects in the factory of the future.

Our purpose in publishing these guidelines is to provide a set of criteria that measure a software architecture's ability to provide an effective starting point and a growth path to full integration of the manufacturing enterprise. Prospective customers and users of automation software can use the guidelines to compare products and to measure how well these products support the requirements of enterprise–wide integration. These guidelines also represent a challenge to the software industry to provide the foundations on which truly integrated enterprises can be built. Until this challenge is met, the real potential of factory automation will never be realized.

2. THE GUIDELINES

Ultimately, the ability of software to support integration over time is a function of the software architecture being employed. The guidelines we have developed to meet one of the INTEGRA objectives are presented as seven questions to ask your vendor or project team about the software architecture of an automation project.

Does the software architecture

- integrate manufacturing with the enterprise?
- support am incremental approach?

- recognize the constant change in people, processes, procedures, and equipment?
- recognize the specific requirements of the manufacturing environment?
- support a distributed environment?
- provide a vendor–independent, open system philosophy?
- anticipate technological trends?

Requirements and features associated with each of these guidelines are discussed in detail in the following guidelines.

2.1. DOES THE SOFTWARE ARCHITECTURE INTEGRATE MANUFACTURING WITH THE ENTERPRISE?

Industry's utilization of manufacturing automation has been fragmented. Some have focused on a top–down approach, enhancing business planning systems with shop floor control features (e.g. on–line WIP tracking and data collection systems). Others have taken a bottom–up approach, building islands of automation (e.g. with PC–based cell data collection and monitoring systems). Still others have focused on the lateral approach of improving CAD/CAE connectivity to manufacturing (e.g. through parts program download links).

Unfortunately, the choice of any one of these focuses results in a local optimization – finding the best solution to a problem at the expense of the overall enterprise. As subsequent automation projects are undertaken, components become isolated, and artificial information barriers are constructed, impeding the flow of data through the enterprise. Solutions to problems in one project carry the risk of adversely impacting problem–solving in future projects. Attempts to tie together fragmented, locally optimized solutions are cost prohibitive, and often unfeasible. The challenge is to overcome this pocket approach to automation and attain full integration of the enterprise.

REQUIREMENTS:
The day of "local optimization", isolated decision–making, and individual problem–solving has long past. Foresight is the key word now. Before applications are built, the manufacturing enterprise software architecture platform must be considered.

The software architecture must be able to span all levels of the plant, enabling data to be readily shared by different users with different views of that data. A fully featured software architecture must:

- download master schedules and material requirements to area–level systems.
- feed shop floor data encompassing orders and delivery requirements back to business systems.
- support area–level systems concerned with scheduling and optimization of resource utilization to meet these requirements.
- Provide cell–level systems that coordinate shop floor devices to implement schedule execution and data collection.
- Provide connectivity to a variety of shop floor device controllers.
- Enable CAD/CAE connectivity to shop floor device controllers for parts program download.

These requirements imply a software architecture allowing data to be accessed among multiple computers and applications to be distributed among several computers. Since every enterprise is different, an architecture that spans all these requirements must be flexible enough to support tailoring to each customer's individual need.

FEATURES TO LOOK FOR:
The software architecture should be constructed so that it can provide a distributed software system supporting area and cell functions and provide smooth integration to business planning and CAD/CAE systems.

In order to lower the cost and risk of automation while providing solutions that fit your enterprise, the architecture should provide a mix of standard products, configurable applications, and development tools. Standard products include run–time platform services allowing applications to be distributed across multiple computers using popular networks. these services should provide guaranteed delivery of data to protect valuable shop floor information.

Libraries of standard shop floor device controller drivers should be provided. These should support a popular mix of PLC's (Programmable Logic Controllers) and CNC'S (Computer Numerical Controllers), and provide an evolutionary path to future MMS (Manufacturing Message Specification within MAP 3.0 Standard Specification) controllers.

Because nearly every automation project has some unique requirements, the architecture should provide a set of development tools that enable the project engineer or systems integrator to do that small percentage of incremental software development distinguishing a great solution that is optimized to the enterprise's critical success factors from a mediocre one. Development tools should be well supported with documentation and training. The software vendor should be able to refer you to a number of qualified software integrators.

Finally, the architecture should provide gateways to the business planning (e.g., Manufacturing Resource Planning) and product engineering (e.g., Computer Aided Design) worlds.

2.2. DOES THE SOFTWARE ARCHITECTURE SUPPORT AN INCREMENTAL APPROACH?

In the past, some large industrial companies rolled the dice and built green field computer–integrated manufacturing plants. The strategy was to invest large amounts of money, put everything under one roof, and push advanced technology in a number of areas.

Today the approach has shifted to implementation of industrial automation projects incrementally as needs and finances dictate. Not every enterprise requires all the features of a full–blown system. Not every enterprise is willing or able to entertain the associated cost/risk burdens. Many enterprises prefer, and in fact must, begin their automation efforts in a focused area. They must have the capability to start the process where they choose, with solutions sized to their environment and needs.

The challenge is how to implement incremental growth while keeping the door open to evolution in the direction of a fully integrated enterprise.

REQUIREMENTS:

In order to support an incremental approach without compromise, the software architecture must support modularity in conjunction with the capability to support area and cell–level integration with business and engineering functions.

In this context, modularity means that the software architecture must offer a growable and scalable approach. It must allow for a user to select and use portions of the architecture independently, with the confidence that the building blocks will eventually fit together into an integrated whole.

FEATURES TO LOOK FOR:
The software architecture should offer good modularity. It should have a base–plus–options packaging scheme, so that the user can tailor functionality to needs. The architecture should support integration of applications from smaller to larger computers, and from single to multiple computers, as requirements grow. It should provide upward compatibility so that automation projects completed one year can be integrated with future products using later, more capable releases of application software. The software architecture must accommodate application upgrades when desired by the user, and support bottom–up and top–down implementation strategies. Finally, the software vendor ought to be positioned to support a mix of approaches to getting the job done: off–the–shelf software products, a roster of system integrators qualified to integrate applications, and training and support for in–house staffs planning customization and doing system maintenance.

2.3. DOES THE SOFTWARE ARCHITECTURE RECOGNIZE THE CONSTANT CHANGES IN PEOPLE, PROCESSES, PROCEDURES, AND EQUIPMENT?

Most business enterprise are constantly changing, and the technology surrounding and supporting them is often changing even faster. The enterprise expands. New products and processes are introduced. Foreign markets are entered, and international manufacturing facilities are set up. Manufacturing plants change and change again. The workforce expands, contracts, changes.

Fierce competition spurs process changes to improve quality and market edge. Equipment is replaced. New equipment is introduced. New operational procedures are initiated. Some of these changes are short–term in nature (e.g., a change in a process variable or a PLC program to improve product quality). Some are longer–term (e.g., addition of a second cell or a new piece of process equipment). The challenge is how to accommodate change, implement it, and take advantage of it without impacting upon the enterprise operation, its capabilities, or the way it conducts its business.

REQUIREMENTS:
The software architecture must be able to respond to internal and external change when it happens. It must accommodate changes in a real–time, on–line manner. The software must be flexible, and must provide functionality that permits on–line reconfiguration of user interface displays and input/output assignments, and/or the addition of new devices to the plant floor. Configurability without the need for programming is the key.

FEATURES TO LOOK FOR:

Configuration tools that allow on–line addition, modification, or deletion of input/output assignments to the process equipment should be available. User displays of graphics or dynamic data should be modifiable or definable in real time and on–line. The user interface and tool set must be tailorable to accommodate the diverse requirements of various levels of management and workers within the enterprise. As plant floor equipment is added, modified, or removed, the software must support new device connectivity options in a configurable manner.

The software should support translation of screens, help messages, and data logs to native languages without modification of application code. When programming is required, tools to manage modifications to software with minimal impact should be available. These include program development and management tools to support the development, maintenance, and tracking of programs; test simulation facilities to allow testing and phased implementation of automation projects; and debug and trace facilities to be used in system development, monitoring, and trouble–shooting.

Finally, the supplier should be committed to a stable application programming interface and upward compatibility, and the same Application Programming Interface (API) should be used for today's PLC's and CNC's, and tomorrow's Manufacturing Automation Protocol (MAP) devices, so that applications will not be affected as next–generation device controllers become available.

2.4. DOES THE SOFTWARE ARCHITECTURE RECOGNIZE THE SPECIFIC REQUIREMENTS OF THE MANUFACTURING ENVIRONMENT?

The manufacturing component of the enterprise operates in a critically–competitive, time–sensitive atmosphere. Delays and downtime are intolerable. Outdated and inaccurate information is unacceptable.

Perhaps no sector of commerce is more demanding of the automation that supports it. The factory floor is one of constant motion. Its perspective is pragmatic: get the job done now. The factory simply will not suffer time lost waiting for computer failure and recovery.

The challenges are myriad: how to track events in a recoverable way; how to ensure that information critical to management doesn't disappear during a computer failure; how to ensure that the system is able to tolerate a hardware failure without impacting upon production and inventory records.

The critical real–time sensitivity of these industrial applications mandate that the cost of automatic recoverability cannot exceed a manual solution method.

REQUIREMENTS:
Recoverability is a key requirement. The software architecture and applications must map external real–time events into internal transactions within the system. These transactions must be recoverable so that significant events are not lost or counted twice, or orders incorrectly executed.

The architecture must provide a range of time–based transactions to satisfy the needs and perspectives of both short–cycle events (e.g., part completion) and longer–cycle events (e.g., batch completions). It must provide the mechanisms to roll the short–cycle events into longer–cycle events, and eventually into updates across gateways (e.g., work–order completion reports to MRP).

The software architecture must be failure–tolerant. If one node in a distributed system fails, the other nodes must continue to operate.

FEATURES TO LOOK FOR:
The software architecture should provide networked program–to–program transaction processing. The architecture should have recovery and restart capability built in, not added on as an after–thought. Transactions should support guaranteed delivery when required, and synchronization of transaction processing with application data bases. Synchronization assures that unless an event is processed successfully, the database will not be updated. The architecture should allow a short–cycle application (e.g., cell control) to continue running and sending recoverable transactions to an area–level function while the area–level function is down.

2.5. DOES THE SOFTWARE ARCHITECTURE SUPPORT A DISTRIBUTE ENVIRONMENT?

Given the hierarchical nature of manufacturing, the issue becomes how to effectively distribute applications across a group of networked computers. Even if the application does not initially need to be distributed, as the enterprise grows, its computer resource needs grow, eventually leading to a distributed manufacturing computing environment.

REQUIREMENTS:
The software architecture should provide flexibility. Without change to the application, the software architecture should allow the enterprise to satisfy

growth by moving up to larger machines and distributing the application among multiple machines.

FEATURES TO LOOK FOR:
The software architecture should provide a methodology for designing, developing, delivering, and maintaining hierarchical, distributed systems. The methodology should allow an application to be decomposed into a number of programs that can be moved around in a distributed computing environment as requirements change. Gateways should be provided to allow the integration of the architecture with business and engineering systems across networked environments.

2.6. DOES THE SOFTWARE ARCHITECTURE PROVIDE A VENDOR-INDEPENDENT, OPEN-SYSTEM PHILOSOPHY?

The locked-in, proprietary system approach was once a compelling argument: select a vendor, and you will live happily forever, fully integrated. Unfortunately, locked-in users are often unable to take advantage of the best price/performance solutions available for a specific application. Also, each computer hardware vendor has areas in which their hardware and software offerings are more or less attractive. The challenge is to provide industrial automation users with the ability to select their best computer hardware solution without endangering the concept of the integrated enterprise model.

REQUIREMENTS:
The software architecture must be able to support a variety of hardware operating systems. It must also recognize the emerging "open" system standards.

The software architecture must support the choice of the best price/performance solutions while protecting the mandate of the integrated enterprise. It must allow the enterprise to install computers from a mix of vendors, and to install the latest technology from each vendor.

FEATURES TO LOOK FOR:
The software architecture should isolate the manufacturing application software from specific hardware and operating systems such as VMS, UNIX, and OS/2, allowing the same applications to run in all three environments.

It should be compatible with leading database products in the manufacturing area, such as Oracle and INGRES. It should run transparently on popular manufacturing networks such as Ethernet, DECnet, and MAP. The architecture should provide drivers for a range of user terminals, including standard alpha–numeric terminals and X–windows devices. Finally, it should provide a standard look–and–feel for user screens, help, and error management.

2.7. DOES THE SOFTWARE ARCHITECTURE ANTICIPATE TECHNOLOGICAL TRENDS?

What was leading–edge technology five years ago is now history. Technological breakthroughts are altering the approaches to problem–solving. The trends are apparent: computer technology is advancing faster than users can absorb. Radically new hardware promises to arrive while current hardware is still in use. Standards for communication and operator interface are rapidly emerging.

How can the enterprise respond to this technological volatility? How can it prepare for the future while still taking advantages of today's leading edge technology?

REQUIREMENTS:
The software architecture should anticipate technological trends allowing today's applications to benefit from future developments. Trends in software development, distributed computing, network technology, device communications, and user interfaces should all be addressed.

FEATURES TO LOOK FOR:
The architecture should provide object–oriented development tools to speed software development and increase software reusability, and provide a natural growth path to a higher generation software technology as it evolves.

It should support distribution of an application across multiple computers.

It should support current proprietary network technology, as well as trends toward open network standards (OSI).

It should support today's devices and protocols (e.g., PLC's) with a device–independent protocol that will transparently support standard protocols as they emerge (e.g., MAP MMS).

It should support the trend toward X–windows terminals, while providing backward compatibility with alpha–numeric terminals.

It should support application software making it transparent to the constantly emerging new hardware and operating system technologies.

3. SUMMARY

When we examine the future of the manufacturing automation industry, it is clear that change in technology is inevitable. We also see that customers look for insurance against technological uncertainty and want to keep their options open. Finally, technological change presents business opportunities for companies that implement new technologies.

There are thus many opportunities for those companies who are willing to bring manufacturing automation on a common international basis. The INTEGRA initiative is part of that effort.

In order to meet these challenges, manufacturing automation software suppliers need to develop products that leverage new technology without precluding full integration of the enterprise.

Suppliers need therefore to innovate to meet customers' needs now and in the future, and adhere to a set of guidelines that allows for change, provide for new technology, and ensure against technological uncertainty.

The INTEGRA objective is to spread such guidelines, so that software development can be standardized. MainStream, MSC's application integration platform for open factory systems, is meeting these requirements by providing a comprehensive tool set for developing, debugging and installing CIM applications at various levels. It embodies three key aspects, conceptually distinct, but all fundamental to a software product aimed at tying traditional business planning and management systems (e.g. MRP) vertically with shop floor control systems:

Chapter 11

HOW TO IMPLEMENT CIM

Joseph SHAMIR
Federico PISANI
ITP AUTOMAZIONE S.P.A.
Via Spalato 7
I-10141 Torino, Italy

Abstract: The purpose of this paper is to fulfil the INTEGRA objective to propose some basic concepts of a Computer Integrated Manufacturing project implementation method. The proposal is based on over a decade of ITP group experience with CIM projects, complemented with the comments of other INTEGRA members. This document defines the building blocks used in this method and discusses the correct approach to the implementation phases and to organizing a CIM System project.

1. INTRODUCTION

Computer technology has progressively pervaded the principal functions of manufacturing activities, and has made possible information systems that integrate not only accounting, administration and commercial aspects but also production and engineering processes.

This production–oriented, integrated information system has been defined as Computer Integrated Manufacturing (CIM). The traditional integrated information system of the 1970's principally involved the integration of organization, accounting, administration, commercial, and marketing aspects. CIM, on the other hand, integrates, as well, processes related to the product (i.e., engineering, shop floor, logistics) and, hence, manufacturing operations activities.

Consequently, CIM can have a strong impact on:

- Cost structure, or the value added structure of the product
- Inventories and their management
- External logistics

- Shop floor organization, including process, material handling, layout, and control.

In other words, CIM can stimulate and support a new manufacturing setup (NMS). NMS is the main leverage of CIM: in addition to saving labour through operation control (normally related to process automation), it can offer substantial operation and capital savings. The actual cost of CIM technology generates unacceptable ROI values unless the benefits of the new manufacturing setup are included.

As a consequence, companies that adopt the CIM leverage actually face two projects simultaneously. They must develop:

1. An integrated information system for operations
2. A new manufacturing setup based on CIM technology

Together, these two projects constitute a "CIM system project".

If a company wants to overcome the obstacle of the cost of CIM technology and confront the organization's natural resistance to change, strong management support and the pressure of competition are necessary.

It is very likely that the future CIM cost will drop and will offer better ROI to companies who adopt a "slow" strategy: nonetheless, organizational barriers and competitive edge relate to the NMS, not the technology. Failure to acknowledge this truth can create a fatal delay in a company's ability to respond to competition, a delay which can have a greater impact than any delay in confronting technological change.

The following sections describe the joint considerations of CIM implementation methods and an integrated information system and step through the NMS approach.

2. THE CIM SYSTEM BUILDING BLOCK

The manufacturing business is a complex "System", which employs operative objectives to drive its organization to perform and control design, fabrication, and logistic activities.

CIM technologies can be classified into:

- Factory automation, local networks, special industrial terminals, shop floor computers, PLC, CNC, DNC, FMB, material storage, transport systems, expert systems, factory control software.

- Engineering automation, personal computers, low cost graphic workstations, CAE, CAD, CAM, CAT, CAPP, word processing, technical publishing, productivity tools, expert systems.

- Traditional information systems: mainframe, business computers, WAN, LAN, MRP, accounting applications, MIS, expert systems, information centers, office automation.

Constructing a CIM system for a particular business requires selecting the technologies most suited for an improved manufacturing setup. These technologies should be mapped onto the "grid" of operation activities and responsibilities in a way that defines "self–governing operative units". These interrelated units are characterized by a well defined autonomy of operation and the management of local resources, including equipment, the information system, and logistics.

The self–governing operative unit of the system are called subsystems. If the subsystems are too complex, the process can be simplified by reshaping the subsystem into more detailed elements, modules, that posses the same self–governing characteristics. Of course, the self–governing autonomy decreases with the increase of details.

The design process phase, which identifies the subsystems, and, eventually, the modules can be called the Integrated System Design (ISD) phase of the CIM system project. This phase is further described in paragraph five.

3. REALIZATION OF A CIM SYSTEM

An integrated system of this dimension (greater than 100 modules) can be faced only through a top down approach. The approach would be similar, then, to the approach used for a large construction project. The CIM system can be described as a "tree structure" because the top down *design phase ensures* that each level considers the integration and coherence of subsequent levels.

The ISD is the conceptual design phase of CIM system and defines its subsystems and modules (nodes of the tree), the system operating model (nodes interfaces), the realization plan, and the economic valuation of the project (Table 6, Phase 1).

The ISD is followed by the detailed design of each subsystem and its development and integration. During this phases the global view of the system, assured by the ISD, guarantees integration without restraining the development of each subsystem.

The startup of the system from the availability of all the parts of the system are progressively tested and integrated, bottom up, until the release of the complete system.

The project development scheme, which tries, through a gradual process to pass from the general business objectives to the detailed realization, represents, in our point of view, the success elements for those projects whose working and organizational aspects must be integrated with information aspects and with complex technological and manufacturing problems.

Furthermore, it is important to define a global program (master plan) thereby the scheduling of activity incorporates each stage of functionality (operating stage) and manages transitional change.

The implementation of a CIM System through a number predefined stages of functionality reduces risks and ensures:

Prior realization of more beneficial subsystems
A soft "evolution" of the organization
A gradual upgrading of the system, which is particularly important
to maintaining ongoing operation through the project

– A uniform distribution of the resources required for the project

The complexity and duration of project activities must be supported by System Integration activity, to ensure the technical co-ordination and coherence of the project choices with the ISD results. The System Integrator has the following tasks:

- Development and updating of the working plan
- Technical auditing of the functional specifications for each project activity

- Control of the integration requirements and of the interface between project activities
- Coordination of the project interfunctional groups
- Technical support for the selection of strategical suppliers
- Project development control and management reporting
- Definition of the starting plan and plant testing
- Management and control of modifications, i.e, priority scheduling and conflict resolving.

A more detailed description of the ISD will follow, as well as a description of the Subsystem Detailed Design (SSD) phase, and the starting and testing phase of the new manufacturing setup. The development phase (Phase 3) i5 described briefly. It includes the following parts that will be developed through traditional methods:

- Hardware and software control systems
- Working development and training
- Installation development and resource and tool acquisition.

4. INTEGRATED SYSTEM DESIGN

The ISD, as already mentioned, is the CIM system concept phase. It starts with data collection for the current manufacturing setup model ("as is" model) and for the validation of the manufacturing economic model.

The integration of the current state, based on comparison with reference data (value) of similar production, identifies the strengths and weaknesses of the business. This data, together with the business objectives, define the new manufacturing setup (NMS) and the related integrated information system (IIS) that will be the CIM System. The objectives of the NMS are verified through the simulation of an economic model of the manufacturing activities.

Once the NMS is defined, the subsystems of the CIM system, and their functional specifications, are identified. Then, the stages of functionality through which the NMS must evolve are identified, as well as the realization plan (master plan) and an appropriate cost/benefit analysis.

5. SUBSYSTEM DETAILED DESIGN

The SDD covers all the subsystems defined in the ISD phase. These subsystems cover all the operations. In particular, if we consider the shop floor, the activities can be classified as follows:

1. Detail definition of the layout

- Cells, machines within the cells, and their allocation in the factory
- Physical interface between cells, as regards primary flow (i.e. materials and rough materials in general) the secondary flow (i.e., equipment and tools), and the relative means of communication and transport.

2. Detailed definition of the subsystem control functions

- Interface function with the integrated system
- Interface functions between cell control systems
- Main elaboration functions at cell and subsystem level
- Main database and localization.

3. Subsystem organization and job descriptions.

4. Definition of the results and of the synchronism between control subsystems.

5. Definition of the subsystem development plan.

In parallel to the detailed design of the subsystem, another activity of this phase should specify the detailed interface of the subsystem and control the coherence of the subsystem. This activity is normally covered by the System Integrator.

6. STARTUP AND TESTING OF THE CIM SYSTEM

This phase covers the actual integration of the subsystems and the global system test. Referring to the shop floor example, this phase includes:

- Gradual startup of the single machine and successful integration of the cell. Job floor personnel are trained to perform cell operations in manual mode.
- Integration with the cell control system. Shop floor personnel are trained to perform cell operations in "semi–automatic" mode.
- Startup of the whole subsystem. Shop floor personnel are trained to perform in full execution mode (no scheduling).
- Full integration of the subsystem into the global information system. Shop floor personnel training includes scheduling for fully integrated mode.

Substantial support is required to ensure a successful System Integration phase, and this support includes clear definition and control of a system startup and test plan.

7. PROJECT ORGANIZATION

A CIM system project demands a highly qualified project manager. The strategic implementation of the project requires that the project manager be a part of and directly supported by top management. In addition, the technical complexity implies a substantial technical coordination effort that, due to the expertise and time required, cannot be performed by the project manager. Therefore the project manager must also be supported by a technically strong System Integrator.

Project activities interfere with and modify the day to day activities of an organization. It i5 crucial that the same line managers responsible for day to day operations be involved in project activities. Their involvement ensures a smooth hand–over of the implemented subsystems.

We recommend the following organizational considerations:

- Subsystem project leaders should be dedicated from the line so they can act as representatives of the innovations during the hand–over phase.

- Line managers should participate on a staff committee, or CIM operative committee, and assist the project manager in planning, and resolving conflicts.

The biggest enemies of a CIM system project are the lack of commitment and delegation on the part of top management, and organizational resistance to change. Communication, training, and the participation of the entire organization, as well as technical capabilities and independence of Software Integrator, are the key aspects of a successful CIM System project.

8. CONCLUSION

A CIM System is based on:

1. An integrated information system for production

2. A new manufacturing setup

CIM System implementation requires activities involving:

- Information systems
- Process, automation and technology
- Logistics and layout
- Organization

The conceptual design of the system (ISD) and its implementation by stage of functionality minimizes risks and permits a gradual evolving of the manufacturing setup.

The project manager's relationship with top management and with the line managers, the support he receives from a technically strong System Integrator, and his ability to introduce change to the entire organization will determine the success of a CIM System project.

Chapter 12

THE FUTURE: AN INTERNATIONAL STANDARD FOR MANUFACTURING SOFTWARE

Dr. Jan–Olaf Willums
Nordic Manufacturing Forum
P.O.Box 301
N–1324 Lysaker, Norway

Abstract: In this final chapter, the present trends are discussed and expanded to explore the possible developments in the future. The link to other European initiatives, including the ESPIT AMICE programme, is made, and suggestions for further developments are outlined.

1. WHERE DO THE TRENDS IN MANUFACTURING LEAD US?

Many people wonder in which direction the CIM trend will develop. In Japan and Germany, we see a trend towards extending the CIM concept to encompass more and more activities within the enterprise. In the Nordic countries, the corporate involvement in CIM has been more moderate, and many observers had to revise their predictions on the growth of the CIM market. It may reflect the economic recession that hit the Scandinavian markets especially hard. But it may also be a sign that too progressive manufacturing technologies have difficulties in penetrating the market: Integrating foreign production philosophies into one's own mental framework of business turned out to be more difficult that many imagined.

The overall goal of the INTEGRA project, and of the Nordic Manufacturing Forum in general, was to exchange the experience among different production facilities and corporate cultures, especially in regards to manufacturing technologies. INTEGRA's objective was to reach some guidelines or "benchmarks" that could guide the further development of CIM in the Nordic countries.

Exchanging views and setting joint priorities will lead to better integration: By bringing the developers of CIM software and the end users in the factories together, many problems of CIM implementation are overcome more effectively.

The understanding of the usefulness of "production platforms" and their integration as a basis for all software developments in a company, will yield efficiencies that are very hard to measure. But experience from other countries indicate that a platform will rapidly reduce the software development cost, and – most importantly – increase the quality of CIM programming and implementation of modern manufacturing technologies.

The INTEGRA initiative started a broader debate about CIM and its impact on industry. The activity in the Nordic Manufacturing Forum expanded the view of Scandinavian manufacturers to include in the CIM concept the aspects of a corporation that are not normally covered by the pure data–approach.

2. THE ESPRIT CONSORTIUM AMICE

The INTEGRA Group expanded therefore the contacts with other programmes in Europe, North America and Japan, and especially with the AMICE project of the ESPRIT Programme. In discussion with manufacturing experts from other countries, we found that we share the concerns we raised when we studied the problems and pitfalls of introducing modern manufacturing concepts in our local factories and enterprises.

The AMICE project has been of special interest to INTEGRA members, and the Nordic Manufacturing Forum arranged as early as 1989 meetings with members of this group, among others through a joint seminar arranged with Siemens in their research center in Munich.

AMICE is part of the European Strategic Programme for Research and Development in Information Technologies (ESPRIT) and is therefore a joint European initiative. AMICE addresses the problem area of CIM system design and implementation. It enjoys wide representative support with the participation of 19 industrial and academic organisations. Its objectives are, broadly speaking, to devise and publicise a CIM Open Systems Architecture which is valid for all types of applications and all types of industries, and which will comply with evolving technologies. The main objectives of the project can be summarised as follows :

- to enable fast, economic utilisation of advanced technologies in industry
- to ensure long range, evolutionary CIM implementation and growth

- to enable and support independent development of CIM building blocks

These objectives are thus very complementary with the INTEGRA initiative, and a combination of INTEGRA efforts and AMICE plans were therefore part of ongoing discussions, as both groups wanted to meet these objectives by:

- establishing a common understanding and terminology for CIM
- providing models as a reference for CIM Planning, design, and implementation
- suggesting and promoting relevant standards in mature areas.

The INTEGRA initiative focused on the latter part of these tasks, with the concrete development of recommendations and the possibility of influence an already well–organised commercial effort in producing a manufacturing software platform.

3. THE CHALLENGE FOR THE SMALL AND MEDIUM–SIZE ENTERPRISES

Several of the studies done as part of the INTEGRA initiative, and articles presented in this final report underline the special manufacturing situation in the Nordic countries, which in general has smaller manufacturing sites that in other parts of Europe or Japan and North America. But even in countries like Germany and the UK, a major part of a nations competitive industry, and especially the manufacturing sector, is dependent on an industrial infrastructure based on efficient and competitive mediumsized enterprises.

For CIM to have a major impact on global manufacturing, it has to be able to adapt to the needs of these small and medium.–size enterprises. (SME's). We have seen that, if properly handled, the elements of CIM can not only benefit the multinational corporations, but also local manufacturers. But the entry cost for the smaller enterprise is often too high, even if the benefits in yielding higher quality, better cost control and smoother production flow can be clearly shown.

The best way to compensate these barriers is to reduce the entry cost to CIM, and we have therefore strongly suggested that so–called productivity platforms, or a standardisation of CIM elements, should be promoted internationally. It would allow companies to choose from off–the–shelf elements for solving the most commonly encountered manufacturing automation challenges. And it would create a market for competing software developers, offering compatible solutions to these manufacturing problems. The result would be a reduction in cost, similar – although maybe not as dramatic– as we have seen recently in the computer hardware sector, and more and more also in the software sector. The manufacturing industry would be the main beneficiary of such a development.

International standardisation is required, however, in order to achieve this objective. We need to see a few platform standards, preferably only one per sector, so that developers can focus their software work on these standards. For that we need an international cooperation among companies and developers in different countries. The INTEGRA project was the first step in that direction, bringing the Scandinavian and European points of view into the discussion. We need now to expand these efforts on a wider basis. The cooperation between INTEGRA members with other EUREKA projects, and the expansions of the cooperation with CIM OSA, and between these European initiatives and North American and Japanese efforts, will at the end benefit the manufacturing industry as a whole. We hope that the INTEGRA project has made its contribution to that effort.

APPENDIX:

INTEGRA – PROJECT APPLICATIONS: A TECHNICAL OVERVIEW OF MAINSTREAM

Prepared for INTEGRA by
Manufacturing Software Development
Via Alpi Graie, 8/B
10095 RIVOLI – TORINO
ITALY
(11) 95.61.700

Abstract: This appendix describes MainStream, a set of software tools, utilities, and applications designed for use in the development of industrial automation applications. It is the first tools that conforms with, and exemplifies the INTEGRA project recommendations. MainStream enables the user to write and configure applications to answer your special industrial automation requirements. It allows to allocate resources for complex configurations and to guarantee the integrity of data. It helps make an industrial application programs recoverable and robust.

1. GENERAL COMMENTS

MainStream consists of the MainStream Base and MainStream Configurable Cell Applications (CCA).

The MainStream Base is designed mainly for area–level industrial automation. It provides a set of software tools that enable you to write event–driven industrial application programs that:

- Are able to run on different types of computer hardware in a distributed environment
- Communicate with other MainStream application programs
- Communicate with a variety of programmable devices, supporting device–specific communications protocols

- Can be developed using a combination of user–configurable tools and C programs

The MainStream Base consists of:

- Libraries of functions that your applications can access by means of a callable programming interface
- Interactive utilities for application development and debugging
- Ready–to–tun applications for logging events, managing alarms, and creating simulated conditions for program testing

MainStream Configurable Cell Applications (CCA) provide userconfigurable tools that make it easy for you to develop applications that support monitoring, alarm setting, and data–tracking functions. These functions can then oversee processes, machines, and the devices that control them. For a number of specialized tasks, CCA provides optional applications, or packages.

MainStream helps insulate your applications from your operating system, making your application code more portable. Both your user–written MainStream applications and your MainStream Configurable Cell Applications access most services of your operating system through MainStream.

The following chapters provide a brief description of MainStream Base software.

2. MAINSTREAM BASE CLUSTERS

MainStream Base services are provided by components known as clusters. Clusters provide a variety of services that support factory automation applications. This section provides an overview of the MainStream Base Clusters.

Each MainStream Base Cluster consists of one or more of the following:

- A MainStream object that the cluster uses to encapsulate information.
- A set of related functions that you can call from a C–language application program. The functions in a cluster perform operations on the MainStream object or objects associated with the cluster. Most

clusters include standard macros, constants, and type definitions that you use in conjunction with the functions. For each Cluster this section will provide a brief description of the related callable interface (A.P.I.: Application Program Interface).

- One or more interactive utilities. MainStream's interactive utilities enable you to define, control, and interact with the programming environment in which your MainStream application program is to run.

2.1. SYSTEM CLUSTER (ISYS)

The System Cluster (ISYS) provides a number of functions that a MainStream program must use in place of standard C run-time library functions, such as studio functions. The MainStream functions are designed to recover if an interrupt occurs while they are executing; the equivalent studio functions cannot recover. Most ISYS functions are identical in other respects to the corresponding studio functions. To help make code portable, MainStream application programmers should use ISYS functions rather than the corresponding studio functions. Some ISYS functions provide functionality not available through standard C run-time functions.

2.2. EVENT MANAGEMENT CLUSTER (IEVT)

The Event Management Cluster (IEVT) enables your MainStream application program to define significant runtime occurrences as events. Using IEVT functions, your MainStream application program can then manage events that it requests, requires knowledge of, or must respond to.

In particular, IEVT enables your MainStream application program to:

- Coordinate events that occur within the context of a single application process
- Manage events synchronously or asynchronously
- Resynchronize the operation of asynchronous MainStream services, such as those provided by the IDSM, ITRN, IMMS, and ITIM MainStream Clusters
- Wait for the completion of a certain event or group of events
- Place completed events from different MainStream Base Clusters in a single queue to control the processing of completed events

- Process completed events at the interrupt level
- Return values to an application program from events that request data
- Check the completion status of each event
- Disable the processing of specified classes of events during sensitive operations.

The Event Management Cluster consists of a library of callable functions that provide applications with access to IEVT services.

Destroy event objects

ievt_destroy() destroys an event object

Specify event action type

ievt_action_dispatch() specifies event action type "dispatch"

ievt_action_queue() specifies event action type "queue"

Set and get event information

ievt_get_class() gets the class attribute associated with an event

ievt_get_error_code() gets the active message code of the MainStream Error Cluster message attribute associated with an event

ievt_get_error_msg() gets the MainStream Error Cluster message attribute associated with an event

ievt_get_result_object() gets the result object attribute associated with an event ievt_get_type() gets the type attribute associated with an event

ievt_get_user_object() gets the user object attribute associated with an event

Event queue management

ievt_queue_create() creates an event completion queue

ievt_queue_destroy() destroys an event completion queue

ievt_queue_is_empty() queries whether an event completion queue is empty

Block or unblock execution of event actions

ievt_manager_block()   blocks execution of event actions for one or more
                                        classes of event

ievt_manager_unblock()   unblocks execution of event actions for one or more
                                        specified classes

Create, add to, or delete event class masks

ievt_mask_create()   creates an event class mask object

ievt_mask_add_class()   adds a class to an existing event class mask

ievt_mask_delete_class()   deletes a class from an event class mask object

ievt_mask_destroy()   destroys the specified event class ask object

Wait for event or lists of events to be issued

ievt_wait()

ievt_wait_list()

ievt_manager_wait()   waits until the event manager detects that an event
                                        is issued

ievt_manager_wait_if()   waits until an event is issued that meets the
                                        predicate

Event action types

IEVT_ACTION_NONE   constant specifying an event action type "no
                                        operation" with valid event object

IEVT_ACTION_NO_EVENT   constant specifying an event action type "no
                                        operation" with silent event object

IEVT_ACTION_SYNCHRONOUS   constant specifying an action type of "synchronous"

Event classes

IEVT_CLASS_IDSM   constant specifying the class of events generated by
                                        IDSM functions

IEVT_CLASS_IMMS   constant specifying the class of events generated by
                                        IMMS functions

IEVT_CLASS_ITIM   constant specifying the class of events generated by
                                        ITIM functions

IEVT_CLASS_ITRN   constant specifying the class of events generated by
                                        ITRN functions

Get next event

ievt_queue_next()	gets next event from queue
ievt_queue_next_if()	

Query

ievt_manager_is_dispatching()	queries whether current routine is executing at dispatch level
ievt_queue_is_empty()	queries whether an event completion queue is empty
ievt_is_issued()	queries whether event has been issued

2.3. EXIT HANDLING CLUSTER (IEXH)

The Exit Handling Cluster (IEXH) enables a MainStream application program to provide both for conditional or unexpected program exits and for a standard program exit upon successful completion of the program. The Exit Handling Cluster enables programs to specify the values that they are to return to the operating system when they exit.

The Exit Handling Cluster also enables programs to perform appropriate tasks immediately before they exit. These tasks are performed automatically by user–written functions known as exit handlers. Exit handlers are useful for cleaning up global or persistent objects such as:

- File contents
- Partially completed operations
- External state information
- Any object or data remaining after the termination of the process

The exit handling logic that a MainStream application program defines for itself using the Exit Handling Cluster is independent of the exit handling logic of the operating system.

The Exit Handling Cluster consists of a library of callable functions that provide applications with access to IEXH services.

Execute handlers and exit

iexh_exit()	Executes exit handlers and terminates the calling program

Return value constants

IEXH_SUCCESS · Return value of MainStream application program exiting with success

IEXH_FAILURE · Return value of MainStream application program exiting with failure

Declare and cancel exit handlers

iexh_declare() · Declares an exit handler

iexh_cancel() · Cancels an exit handler

Force exit

iexh_force_exit() · Forces a process to exit

Indicate whether exit in progress

iexh_exit_in_prog() · Indicates whether exiting is in progress

Enable/disable code dump on exit

iexh_enable_debug() · Enable or disable generation of debugging information during exiting.

2.4. ERROR HANDLING CLUSTER (IERR)

The Error Handling Cluster (IERR) provides MainStream application programs with:

- A consistent and structured way to handle runtime errors, independent of your operating system
- A standard way to distinguish different levels of severity among runtime errors
- Error messages that provide information about the location and nature of the error
- Error recovery and continued execution of program
- Support for cleanup of local or transitory objects such as memory. file descriptors, partially completed operations, and internal state information

• Message files in which you can define error messages for your MainStream application program

The Error Handling Cluster consists of a library of callable functions and one or two utility programs, depending on your operating system. The functions enable applications to access IERR functionality. The utility programs enable you to incorporate message files that you have created into the executable version of your MainStream application program.

Error trap management

IERR_END	ends an error trap
IERR_NAMED_TRAP()	begins a named error trap
IERR_TRAP	begins an unnamed error trap

Output error message and exit

ierr_exit()	outputs error message and exits
ierr_exit_list()	outputs multilevel error message and exits

Error handler management

IERR_CONTINUE	resumes execution after signalled error
IERR_HANDLER()	begins unnamed-handler definition
IERR_INIT_HANDLER()	initializes an error handler
IERR_NAMED_HANDLER()	begins named handler definition
IERR_RESIGNAL	directs execution to next handler
IERR_TO_HANDLER	allows data sharing between main-line and errorhandler code

Severity level codes

IERR_ERROR	denote error-level severity
IERR_FATAL	denotes fatal-level severity
IERR_INFO	denotes informational-level severity
IERR_SUCCESS	denotes success-level severity
IERR_WARNING	denotes warning-level severity

Examine or set severity level

ierr_code_get_severity()	obtains severity level from error code variable

ierr_code_get_successful()	checks whether severity of error code denotes successful or unsuccessful
ierr_code_set_severity()	sets severity level of error code variable
IERR_IF_ERROR()	checks if error code variable is error–level severity
IERR_IF_FATAL()	checks if error code variable is fatal –level severity
IERR_IF_INFO()	checks if error code variable is informational–level severity
IERR_IF_SUCCESS()	checks if error code variable is success level
IERR_IF_SUCCESSFUL()	checks if error code variable indicates success
IERR_IF_NOT_SUCCESSFUL()	checks if error code variable indicates failure
IERR_IF_WARNING()	checks if error code variable is warning–level

Examine or set error code values

ierr_code_get_condition_id()	obtain error condition from error code variable
ierr_code_get_facility_number()	obtains facility component from error code variable
ierr_code_get_message_number()	obtains mnemonic attribute of error code variable
ierr_code_get_inhibit()	checks INHIBIT_PRINT flag of error code variable
ierr_code_set_inhibit()	sets INHIBIT_PRINT attribute of error code variable
ierr_msg_get_number_error_codes()	returns the number of levels in a message object
ierr_msg_get_error_code()	returns an error code from a specified level of a message object

Signal an error condition

ierr_errno_signal()	signals error condition after UNIX system call
ierr_errno_stop()	signals error condition after UNIX system call; stops if continued
ierr_errno_translate	returns error code for operating system error integer
ierr_msg_signal()	signals an error condition using a message object
ierr_msg_stop()	signals an error condition using a message object; stops if continued
ierr_signal()	signals an error condition
ierr_signal_list()	signals a multilevel error condition
ierr_insert()	inserts contextual error condition and signals result
ierr_insert_list()	inserts multilevel error condition and signals results
ierr_stop()	signals an error condition; stops if continued

| ierr_stop_list() | signals multilevel error condition; stops if continued |

Print error messages

| ierr_msg_print() | outputs contents of message object |
| ierr_msg_print_string() | writes message object contents to string variable |

Control unwinding

IERR_IF_UNWINDING	begins error handler's unwinding section
IERR_TO_CALLER	specifies path of an unwind
IERR_TO_TRAP	specifies path of an unwind
IERR_UNWIND	begins an unwind operation
IERR_UNWOUND	ends an error handler's unwinding section

2.5. TRACE CLUSTER (ITRC)

The Trace Cluster (ITRC) enables your application program to report diagnostic messages when it is executing. By invoking ITRC form appropriate places in your application program, you enable the program to report diagnostic messages that trace its execution. In particular, ITRC is useful for tracing:

- When functions start and finish executing
- The results of conditional logic
- Values of variables at runtime

The Trace Cluster also provides a Trace Interactive Utility that enables the user to create, modify, and destroy a Trace database. A Trace database determines which functions in an application program can report Trace messages. Thus, you can enable or disabled different combinations of Trace messages at run time.

2.6. LOCK MANAGEMENT CLUSTER (ILCK)

The Lock Management Cluster (ILCK) provides a way to control how processes access shared resources, such as shared memory, files, records, or other data structures. Competition among processes for access to shared resources can corrupt or destroy the data associated with these resources. By regulating this competition, the Lock Management Cluster helps preserve the integrity of shared resources.

In particular, ILCK enables processes to:

- Create lock objects for particular shared resources
- Organize lock objects into a hierarchy that represents the logical relationship of the corresponding shared resources to one another
- Place locks on lock objects to regulate access to the corresponding shared resources
- Remove locks and destroy unneeded lock objects

The Lock Management Cluster consists of a library of callable functions that provide applications with access to ILCK services.

Manage Lock Objects

ilck_create()	creates a root lock object
ilck_create_child()	creates a child lock object
ilck_destroy()	destroys a lock object
ilck_lock()	places a lock on a lock object
ilck_unlock()	removes a lock from a lock object

Return information

ilck_get_count()	returns the number of locks on a lock object at a specified lock level
ilck_get_total()	returns the total number of lock on a lock object

Constants specifying lock levels

ILCK_EXCLUSIVE	constant specifying the Exclusive Access lock level
ILCK_PROTECTED	constant specifying a Protected Access lock level
ILCK_PROTSHARE	constant specifying Protected Share Access
ILCK_SHARE	constant specifying a Share Access lock level
ILCK_NULL	constant specifying a Null Access lock level

2.7. SHARED MEMORY CLUSTER (ISHR)

The Shared Memory Cluster (ISHR) enables two or more processes to access the same block of physical memory in order to exchange information or share data. A MainStream application program can use ISHR functions to manage shared memory and, if desired, IMEM functions to manage this memory. In this way, ISHR and IMEM enable more than one process to access the same shareable partition at the same time.

ISHR is particularly useful for providing common data stores, dynamic screen displays, and interprocess communication.

The Shared Memory Cluster consists of a library of callable functions that provide applications with access to ISHR services.

Create and delete shared memory regions

ishr_create()	creates shared memory
ishr_destroy()	destroys access to shared memory
ishr_get_begin_address()	finds the starting address
ishr_get_end_address()	finds the end address plus one
ishr_remove()	removes shared memory image file

Write shared memory to file

ishr_flush()	writes shared memory to image file

Shared memory access codes

ISHR_READ_ONLY	specifies read–only memory access to shared memory region
ISHR_READ_WRITE	specifies read–write memory access to shared memory region

2.8. MEMORY CLUSTER (IMEM)

The Memory Cluster (IMEM) provides software tools that manage processes' use of memory. IMEM enables MainStream application programs to:

- Allocate and de–allocate objects dynamically within local memory
- Allocate and de–allocate objects dynamically within shared memory (when used in conjunction with ISHR)
- Share the location of these dynamically allocated objects with other processes

The Memory Cluster consists of a library of callable functions that provide applications with access to IMEM services.

Allocate and deallocate memory

imem_allocate()   allocates memory

imem_free()   deallocates memory

Manage memory partitions

imem_close()

imem_create()   creates memory management partition or accesses existing partition

imem_destroy()   destroys either access to a memory partition or the partition itself

imem_open()

Maintain data in directory

imem_directory_add()   adds an entry to the directory

imem_directory_delete()   deletes an entry from the directory

imem_directory_find()   finds an entry in the directory

Control memory tracing

imem_audit()   controls memory tracing

Convert pointer types

imem_absolute()   converts relative to absolute pointer

imem_relative()   converts absolute to relative pointer

Memory access codes

IMEM_READ

IMEM_READ_WRITE   specifies read–write memory access to memory partition

2.9. TIME MANAGEMENT CLUSTER (ITIM)

The Time Management Cluster (ITIM) enables your MainStream application program to access the operating system's time clock and manipulate expressions of time. ITIM isolates your application from your operating system's own time formats.

In particular, ITIM enables MainStream application programs to:

- Perform arithmetic operations on expressions of time (a date and time of day, or an expression of a duration of time in days, hours, minutes, and seconds)
- Compare expressions of time
- Convert expressions of time from one format to another
- Read the current time from the system time clock
- Suspend a process for a specified time
- Set timers used in application logic

Services of the Time Management Cluster are used by the Test Simulator Utility (ITSU) to define a test clock by which time passes more quickly or slowly than conventional time. Test clocks can be useful in the testing of applications that are under development.

The Time Management Cluster consists of a library of callable functions that provide applications with access to ITIM services.

Operations on time expressions

itim_add()	adds two times and returns their sum
itim_cmp()	compares two times and returns result
itim_collate()	collates two delta or two absolute times
itim_max()	returns largest of two or more time values
itim_min()	returns smallest of two or more time values
itim_sub()	substracts time value and returns their difference
itim_verify_abs()	finds out whether a variable contains an absolute time expression
itim_verify_del()	finds out whether a variable contains a delta time expression

Format conversion and value assignment

ITIM_DELTA_FROM_CONSTANT()	assigns a time duration to a delta time variable
ITIM_DELTA_INITIALIZER(	initializes a variable with a delta time
itim_from_ascii()	converts ASCII time format to internal time format
itim_from_native()	converts a native time expression to internal time format

itim_from_real()	converts floating point to internal delta time
itim_get()	returns the current environmental time
itim_sys_get()	converts internal time expression to ASCII format
itim_to_ascii_buffer()	converts internal time expression to ASCII format and returns converted expression to a user specified buffer
itim_to_components()	converts an absolute time expression into separate integer components
itim_to_native()	converts an expression in internal time format to native time format
itim_to_real()	converts internal format time expression to double precision floating-point
itim_translate()	translates an internal format time expression into an ASCII string
itim_translate_buffer()	translates internal format time expression into an ASCII string, and returns a count of the number of bytes in the string

Suspend process for specified time

itim_sleep()	suspends calling process for specified time
itim_sleep_cancel()	starts process suspended by itim_sleep()
itim_sys_sleep()	suspend calling process for a specified time
itim_sys_sleep_cancel()	starts process suspended by itim_sys_sleep()

Access timers

itim_sys_timer_cancel()	cancels timer set by itim_sys_timer_set()
itim_sys_timer_set()	sets a timer to expire at a specified operating system time
itim_timer_cancel()	cancels timer set by itim_timer_set()
itim_timer_set()	sets a timer

2.10. CHAINED RECORD MANAGEMENT CLUSTER (ICRM)

The Chained Record Management Cluster (ICRM) enables a MainStream application program to store data objects known as chained records in multiple doubly-linked lists. You can optionally incorporate chained records into an indexed sequentially–accessed data structure. Chained records are especially useful for storing the data used to manage and schedule area–level resources.

An application can access data in chained records quickly and can view it from different perspectives. An application can create chained records in local or shared memory.

The Chained Record Management Cluster consists of a library of callable functions that enable applications to access ICRM services. You must write ordering functions to define the ordering criteria for the chained records.

Construct Level Functions

icrm_copy()	copies an ICRM data construct
icrm_create()	creates an ICRM data construct or maps to an existing construct
icrm_destroy()	destroys the specified data construct or access to the construct
icrm_get_mem_obj()	returns the IMEM object for the data construct
icrm_get_name()	returns the name of the data construct
icrm_get_num_recs()	returns the total number of data objects in all chains
icrm_get_num_recs_in_chain()	returns the number of data objects in a chain
icrm_get_num_recs_in_class()	returns the number of data objects in a class
icrm_get_user_parm()	returns the user parameter given as an argument to
icrm_sort()	sorts all the chains in an ICRM construct
icrm_sort_chain()	sorts the data objects in a chain
icrm_sort_class()	sorts the chains in a chain class

Chainhead Level Functions

icrm_chainhead_add()	adds a chainhead to a chain class
icrm_chainhead_delete()	deletes a chainhead
icrm_chainhead_find()	finds a specific chainhead
icrm_chainhead_traverse()	traverse a class of chainheads
icrm_chainhead_traverse_init()	initializes a chainhead context object

Record Level Functions

icrm_rec_add()	adds a data object to the data construct

icrm_rec_delete()	deletes a data object from a chain
icrm_rec_find()	finds a data object in the data construct
icrm_rec_is_linked_to_chain()	returns whether a data objectis linked to a chain
icrm_rec_link_at_ctx()	links a data object to a chain before or after a given data object
icrm_rec_link_to_chain_end()	links a data object to one end of a chain
icrm_rec_link_sorted()	links a data object to a chain according to ordering scheme
icrm_rec_traverse()	traverses each data object in a chain
icrm_rec_traverse_init()	returns an initialized record context object
icrm_rec_unlink()	unlinks a data object from a chain

2.11. PRIORITY QUEUE CLUSTER (IPRQ)

The Priority Queue Cluster (IPRQ) enables a MainStream application program to organize data objects into a structure known as a priority queue that provides fast access to important data, such as data needed for the processing of alarms or for other urgent tasks. MainStream supports priority queues in local or shared memory.

IPRQ enables a MainStream application to establish the relative priority of the objects in the list. By assigning a relatively high priority to certain objects in the priority queue, a program ensures that they can be accessed quickly.

The Priority Queue Cluster consists of a library of callable functions that enable application programs to access IPRQ services. You must write an ordering function to compare two data objects and determine their relative priority according to criteria that you define.

Create or Destroy Priority Queue

| iprq_create() | creates a priority queue |
| iprq_destroy() | destroys a priority queue |

Add or Delete Node from Priority Queue

| iprq_add() | adds a node and its associated data object to the queue |
| iprq_delete_current() | deletes a node from a priority queue |

iprq_delete_top()   deletes the node containing the highest priority data object from the queue

Return Information from Priority Queue

iprq_examine_current()   returns data contents of node in priority queue

iprq_examine_top()   returns top priority data object in priority queue

Replace Data Object in Priority Queue

iprq_replace_current()   replaces data object of node in priority queue with new data object

Traverse Priority Queue

iprq_traverse()   traverses one node of priority queue and updates context variables

iprq_traverse_init()   initializes a context variable used to traverse a priority queue

Type Definitions

IPRQ   type definition for priority queue objects

IPRQ_CTX   type definition for context variables

Constants

IPRQ_ACCESS   specifies that only access to shared memory queue be destroyed

IPRQ_ALL   specifies that access to priority queue and queue itself be destroyed

IPRQ_EXISTING   specifies that iprq_create() access existing shared memory queue

IPRQ_EXISTING_OR_NEW   specifies that iprq_create() create a new queue or access an existing queue

IPRQ_FORWARD   specifies forward direction for priority queue operations

IPRQ_NEW   specifies that iprq_create() create a new queue

IPRQ_REVERSE   specifies reverse direction for priority queue operations

2.12. QUEUE CLUSTER (IQUE)

The Queue Cluster (IQUE) enables a MainStream application program to organize data objects into a FIFO (first–in, first–out) queue. A FIFO queue is useful for storing data objects that need to be processed in the order in which they were created. MainStream supports these queues in local or shared memory.

The Queue Cluster consists of a library of functions that an application program can invoke to create a queue, add objects to and remove them from the queue, and walk through the queue in FIFO order.

Type Definitions

IQUE	type definition for FIFO queue objects
IQUE_CTX	type definition for context variables used to traverse FIFO queue

Constants

IQUE_ACCESS	specifies that only access to shared memory queue be destroyed
IQUE_ALL	specifies that access to the queue and the queue itself be destroyed
IQUE_EXISTING	specifies that ique_create() access existing shared memory queue
IQUE_EXISTING_OR_NEW	specifies that ique_create() create new or access existing queue
IQUE_NEW	specifies that ique_create() create a new queue

Create or Destroy a FIFO Queue

ique_create()	creates and initializes a FIFO queue, or maps to an existing queue
ique_destroy()	destroys a FIFO queue or access to a queue

Add or Remove Queues

ique_add()	adds a new tail node and its data object to a FIFO queue
ique_delete()	deletes the head node and its associated data object from a FIFO queue

Concatenate Queues

ique_cat()   concatenates two FIFO queues

Get Information from Queues

ique_examine()   returns the data object that has been in a FIFO queue longest

ique_extract()   extract a data object from a FIFO queue

ique_is_empty()   returns a boolean indicating whether a FIFO queue is empty or not

ique_size()   returns the size of a FIFO queue

Traverse Queue

ique_traverse_init()   initializes context variable used to traverse a FIFO queue

ique_traverse()   traverses a FIFO queue, one node with each call

2.13. SORT CLUSTER (ISRT)

The Sort Cluster (ISRT) enables a MainStream application program to sort data objects in an array. The data objects can be of any valid C data type and can range in complexity from simple integers to complicated data structures.

   The Sort Cluster provides a function that implements the QUICKSORT algorithm, which is used to sort data objects. You write your own ordering function to determine which data object precede others in the sorted output.

Sorting Functions

isrt_quicksort()   sort an array of data objects using user–written ordering function

isrt_set_create()   creates a set object

isrt_set_destroy()   destroys a set object

isrt_set_get_location()   returns memory location of set object

isrt_set_get_size()   returns size of array

isrt_set_set_location()   associates a set object with an arra location

isrt_set_set_size()   passes an array size to a set object

Constants

ISRT_EQUAL	returned by user-written ordering function when compared objects are equal
ISRT_GREATER	returned by ordering function if first of compared objects is larger
ISRT_LESS	returned by ordering function if first of compared object is smaller

Type Definitions

ISRT_SET	type definition for ISRT set objects

3. TRANSACTION MESSAGING

MainStream provides the following facilities to enable your applications to support process–to–process messages known as transaction messages:

- Transaction Messaging Cluster (ITRN)
- Interactive Transaction Utility (ITU)
- Mnemonic Data Dictionary (IMDD)
- Transaction Monitoring Utility (ITMU)

3.1. TRANSACTION MESSAGING CLUSTER (ITRN)

The MainStream Messaging Cluster (ITRN) enables MainStream application programs to communicate with each other by means of transaction messages. A transaction message is a message with a specified format and length. Transaction messages can be sent and received over networks by MainStream application programs on different computers.

ITRN provides an optional recovery facility. Recoverable transaction processing ensures that a transaction message interrupted by a system failure reaches its destination after the system is restarted. Recovery is ensured by storing, in a recovery database, information that relates to each recoverable transaction message that is sent.

With the recovery facility, transaction messages can be:

- Recovered after problems such as loss of power, abnormal program termination, or failure of operating systems, or networks
- Synchronized with application–related database updates (commits)

The Transaction Messaging Cluster consists of a library of callable functions and a utility, the Interactive Transaction Utility (ITU).

Managing processes as transaction processors

ITRN_PROC	type definition of the transaction message processor object
itrn_proc_create()	enables a process to function as a transaction message processor
itrn_proc_destroy()	destroys a transaction message processor object
itrn_proc_initiated()	begins transaction messaging activity in a transaction message processor
itrn_proc_terminate()	ends transaction message processing and rolls back to initial state
itrn_proc_complete()	completes transaction messaging activity

Creating and sending transactions

ITRN	transaction message object type definition
ITRN_CLASS	type definition for the recoverability attributes
ITRN_CREATE_AND_SET_UP()	creates and populates a transaction message with data
itrn_clone()	creates a new copy of a transaction message
itrn_create()	creates and initializes a transaction message
itrn_process_event()	returns the transaction message processor object associated with an IEVT event
itrn_proc_put_trn()	puts a transaction message
itrn_proc_reput_trn()	requeues or redirects a received transaction message
itrn_set_class()	sets the class attribute of a transaction message
itrn_set_code()	sets the code attribute of a transaction message
itrn_set_delay()	sets the delay attribute of a transaction message
itrn_set_destination()	sets the destination attribute of a transaction message

itrn_proc_synchronize()	insures that all transaction messages in output queue are sent (if links exist)
itrn_destroy()	destroys a transaction message object
itrn_get_class()	returns the class attribute of a transaction message
itrn_get_code()	returns the code of a transaction message
itrn_get_created()	returns the environmental time for the transaction message
itrn_get_data()	returns a pointer to the beginning of the data portion of a transaction message
itrn_get_delay()	returns the delay attribute of a transaction message
itrn_get_destination()	returns the destination attribute of a transaction message
itrn_get_length()	returns the length of the data in a transaction message's data portion
itrn_get_original_source()	returns the key of the original sending transaction message processor
itrn_get_reputs()	returns the number of reputs
itrn_proc_get_trn()	gets a transaction message off the input queue
itrn_proc_wait()	waits for a transaction message
itrn_trace()	prints the attributes and data fields of a transaction message
itrn_proc_traverse()	lets you view a transaction message in the input queue without getting it
itrn_proc_traverse_init()	initializes context variable used to traverse transaction message input queue
itrn_proc_unput_trn()	cancel the sending of a delayed transaction message

3.2. INTERACTIVE TRANSACTION UTILITY (ITU)

The Interactive Transaction Utility (ITU) provides commands that enable you to:

- Create, send, and receive transaction messages interactively to prototype or test your implementation of ITRN
- Prepare a recovery environment for transaction message processing
- Create a Mnemonic Data Dictionary (IMDD)

- Enter information necessary for transaction message processing into a Mnemonic Data Dictionary (IMDD) and use this information to format transaction messages
- Create a database used by the Transaction Monitoring Utility (ITMU)
- Invoke the Transaction Monitoring Utility (ITMU) to view statistics about the current activity of your transaction message processors

3.3. MNEMONIC DATA DICTIONARY (IMDD)

A Mnemonic Data Dictionary (IMDD) contains information such as structure definitions for transaction messages and transaction message codes and their mnemonic symbols. A Mnemonic Data Dictionary is used at runtime by the following MainStream utilities:

- Test Simulator Utility (ITSU)
- Event Logger Utility (IELU)
- Alarm Management Utility (IAMU)

IELU and IAMU use the MDD to define the format of dynamic run–time data in logging and alarm messages.

A Mnemonic Data Dictionary makes it possible to:

- Input or view transaction messages by referencing mnemonic symbols for fields in the transaction message data structure
- Use the Test Simulator Utility to convert test events created by simulated processes during the testing of a program into transaction messages that can be understood by the program that you are testing

3.4. TRANSACTION MONITORING UTILITY (ITMU)

From within the Interactive Transaction Utility, you can invoke the Transaction Monitoring Utility (ITMU) to collect and display real–time statistics on your current transaction message processing environment. This information concerns transaction message processors and the links between them. It enables you to gauge the status, balance, efficiency, and level of activity of your transaction message processing environment.

3.5. *DYNAMIC SCREEN MANAGER (IDSM)*

The Dynamic Screen Manager (IDSM) is a library of functions that manage the interface between a MainStream application program and the user who views and enters data on a video display terminal or X–windows display. Callable IDSM functions can define the appearance of terminal windows and specify how data is to be entered and processed.

IDSM provides the MainStream application programmer with the following features:

- Input and output data on screen
- Full configurability of windows, menus, scrolling regions, keyboard, help messages, privileges, and validation
- A variety of menu formats: pop–up, horizontal, or vertical
- A modular approach to control logic and window layout routines
- A keyboard translation table to define keys
- Integrated field validation
- Xterm compatibility
- VT100/200 compatibility

Used in conjunction with IDSM are the Help Cluster and the Security Cluster.

Objects

IDSM	user interface object type definition
IDSM_MENU	menu object type definition
IDSM_ORIGIN	origin object type definition
IDSM_SCROLL	scroll object type definition
IDSM_WINDOW	window object type definition

Display

IDSM_BOLD	specifies display rendition as bold
idsm_clear_window()	clears the window
IDSM_FLASH	specifies display rendition mode as flash
IDSM_NORMAL	specifies display rendition mode as normal
IDSM_REVERSE	specifies display rendition mode as reverse
idsm_set_mode_background()	sets the background display rendition mode

idsm_set_mode_foreground()   sets the foreground display rendition mode

idsm_text_area()   fills a rectangle with blancks

IDSM_UNDERLINE   specifies display rendition mode as underline

Cursor Manipulation

idsm_inc_x_set_y()   sets text cursor position, x relative and y absolute

idsm_inc_xy()   sets text cursor position, x and y relative

idsm_newline()   moves the cursor down to the beginning of a new line

idsm_origin_create()   creates the origin object

idsm_origin_destroy()   destroy the origin object

idsm_set_x_inc_y()   sets text cursor position, x absolute and y relative

idsm_set_xy()   sets text cursor position, x and y absolute

I/O Fields

idsm_field_choice()   defines a choice input field

idsm_field_double()   defines a floating point input field

idsm_field_integer()   defines an integer input field

idsm_field_key()   defines an input field of zero width

idsm_field_string()   defines a string input field

idsm_field_string_invisible()   defines an invisible string input field

idsm_field_string_lowercase()   defines a lowercase string input field

idsm_field_string_uppercase()   defines an uppercase string input field

Output Functions

idsm_date()   returns the current date and time

idsm_print_double()   prints a floating point number

idsm_print_integer()   prints an integer in a window field

idsm_print_string()   prints a string

idsm_print()   prints a formatted string

idsm_static_print_string()   prints a static string

idsm_static_print()   prints a static formatted string

Field Validation Function

idsm_true()   always true validation function

Scrolling Functions

IDSM_SCROLL_BEGIN	begins a window scroll loop
idsm_scroll_created()	creates the scroll object
idsm_scroll_destroy()	destroys the scroll object
IDSM_SCROLL_END	closes a window scroll loop
idsm_scroll_get_current_row()	returns the current scrolling row
idsm_scroll_more_down()	tests if scrolling region can be scrolled down
idsm_scroll_more_up()	tests if scrolling region can be scrolled up
idsm_scroll_set_max()	resets maximum scrolling region rows

Menu Functions

idsm_menu_add_item()	adds item to menu
idsm_menu_add_menu()	adds a secondary menu to menu
idsm_menu_create()	creates the menu object
idsm_menu_destroy()	destroys the menu object
idsm_menu_horizontal()	displays a horizontal menu
idsm_menu_pull_down()	displays pull–down menu
idsm_menu_vertical()	displays vertical menu

Keyboard Constants

IDSM_KEY_BEGIN	moves to the beginning position within a field
IDSM_KEY_BOTTOM	moves to bottom of scrolling region
IDSM_KEY_CANCEL	rejects data and terminates window
IDSM_KEY_CANCEL_ENTRY	sets the current field to its prior value
IDSM_KEY_CLEAR_ENTRY	clears the current field's value
IDSM_KEY_DATA	switches input focus to data fields
IDSM_KEY_DELETE	erases the input character
IDSM_KEY_DOWN	moves 1 row down or moves to previous menu option
IDSM_KEY_END	moves to the last position within a field
IDSM_KEY_ENTER	accepts data and terminates window or executes current menu option
IDSM_KEY_HELP	displays help message
IDSM_KEY_HOME	moves to the first data input field
IDSM_KEY_LAST	terminates current field and goes to the previous field or moves to previous menu option

IDSM_KEY_LEFT	moves to the next character on the left within a field or moves to previous menu option or cycles to previous choice field constant
IDSM_KEY_MENU	switches input focus to menu
IDSM_KEY_NEXT	terminate current field and goes to the next field or moves to next menu option
IDSM_KEY_NEXT_PAGE	moves to next page
IDSM_KEY_NOOP	defines a keyboard key as inoperative
IDSM_KEY_PREV_PAGE	moves to previous page
IDSM_KEY_PRINT	generates a hard copy of the window
IDSM_KEY_REFRESH	clears and repaints the window
IDSM_KEY_RIGHT	moves to the next character on the right within a field or moves to next menu option or cycles to the next choice field item
IDSM_KEY_TIMEOUT	refreshes window because of timeout
IDSM_KEY_TOP	moves to top of scrolling region
IDSM_KEY_UP	moves 1 row up or moves to previous menu option

Security Functions

idsm_check_privilege()	checks user's privileges
idsm_confirm_user()	checks if user has changed
idsm_get_user_id()	returns the ID of current user
idsm_login_layout()	default login layout routine
idsm_new_user()	logs in a new user

Messaging Functions

idsm_message_field()	defines field for message display
idsm_put_field_message()	displays message below field or menu option
idsm_window_put_message()	displays a message

Help Functions

idsm_help_layout_init()	initializes custom help display
idsm_help_sublayout()	default help display layout

Control Functions

idsm_beep()	sounds a bell

idsm_create()	creates the user interface object
idsm_destroy()	destroys the user interface object
idsm_flush_keyboard()	flushes the keyboard buffer
idsm_get_application_name()	returns the application's name
idsm_get_display_type()	returns the display type
idsm_get_key()	gets the next keyboard entry
idsm_window_create()	creates the window object
idsm_window_destroy()	destroys the window object
idsm_window_display()	displays window
idsm_window_display_updated()	updates the display of a window
idsm_window_get_focus()	returns the current input focus
idsm_window_get_menu_item()	returns the current menu item tag
idsm_window_get_name()	returns the window name
idsm_window_hard_copy()	generates screen dump file
idsm_window_process()	processes window input and updates display
idsm_window_process_event()	processes window based on IDSM event
idsm_window_put_message()	displays a message
idsm_window_set_fields()	change allowable data entry fields
idsm_window_set_menu()	changes menu to display and process
idsm_window_set_origin()	sets the location of a window on the screen
idsm_window_set_size()	define window size in characters
idsm_text_box()	draws the outline of a rectangle
idsm_text_line()	draws a horizontal or vertical line

3.6. HELP CLUSTER (IHLP)

The Help Cluster (IHLP) enables you to create and display online help for IDSM screens and input fields. An interactive utility enables you to create and maintain the help files that you display.

3.7. SECURITY CLUSTER (ISEC)

The Security Cluster (ISEC) provides a system of user ID passwords and privileges that enable you to control access to your applications. The Interactive Security File Utility enables you to create and maintain this system.

ISEC passwords enable you to specify which IDSM windows a user can log into. Thus, you can limit access to your application and to windows within a MainStream application program.

In particular, the Interactive Security File Utility enables you to:

- Create and maintain files that establish passwords and privileges
- Specify the names of these files in your application
- Call functions from your application that log in the user, verify that the same user is still logged in, and check the user's privileges

3.8. MANUFACTURING MESSAGE SPECIFICATION CLUSTER (IMMS)

The Manufacturing Message Specification Cluster (IMMS) supports communications between a MainStream application program and both MMS and non–MMS devices. An MMS device is a device that communicates with other devices or processes on a network according to standards set up by the International Standards Organization (ISO). This standard, known as the Manufacturing Message Specification, follows the ISO Reference Model for Open Systems Communication.

The Manufacturing Message Specification Cluster treats all factory–floor devices as if they were MMS devices and enables you to mix MMS and non–MMS devices in your system. MainStream provides device drivers for a number of non–MMS devices.

The Manufacturing Message Specification Cluster consists of a library of callable functions that provide applications with access to MMS services.

Creating, accessing, and destroying device objects

imms_device_create()	creates a device object
imms_device_destroy()	destroys a device object
imms_device_get_assoc_status()	returns the device association status
imms_device_get_info_status()	gets the status of the information reports service on the device
imms_device_get_istatus()	returns the logical status of a device
imms_device_get_model()	returns a pointer to a string containing the device model name
imms_device_get_nth_capability()	returns a pointer to the nth MMS capability in the list of capabilities defined on the device

imms_device_get_pstatus()	returns the physical status of a device
imms_device_get_revision()	returns a pointer to a string containing the device revision
imms_device_get_vendor()	returns a pointer to a string containing the device vendor name
imms_device_param_is_supported()	returns whether a specific conformance building block is supported on a device
imms_device_service_supported()	returns whether a particular MMS service is supported on the device

Creating, accessing, and destroying domain objects

imms_domain_add_capability()	stores a string containing a domain capability
imms_domain_create()	creates an IMMS domain object associated with an IMMS device object
imms_domain_destroy()	destroys a domain object
imms_domain_get_nth_capability()	returns a pointer to a string referring to the nth domain capability
imms_domain_get_state()	returns the current state of a domain
imms_domain_get_uploads()	returns the current number of uploads in progress on the device
imms_domain_lookup_fn()	retrieves a pointer to the filename string for the domain
imms_domain_set_sharable()	identifies whether the domain is sharable

Creating, accessing, and destroying variables

imms_var_create()	creates an IMMS variable object
iimms_var_destroy()	destroys an IMMS variable object
imms_var_is_deletable()	returns whether a variable is deleteable
imms_var_lookup_dl()	looks up a variable's dl–string in the ISCU database
imms_var_get_device_name()	returns the name of the device associated with a particular variable
imms_var_get_size()	returns the size of the specified variable
imms_var_get_status()	returns the status of the variable object
imms_var_get_user_object()	returns the user object associated with the variable object
imms_var_get_value()	returns the value of the variable object

imms_var_get_value_pointer()   returns a pointer to an area in memory where the value of a variable is stored

imms_var_set_user_object()   sets the user object associated with a variable object

imms_var_write_list()   writes a variable list to the specified device

Creating, accessing, and destroying variable lists

imms_var_list_create()   creates an IMMS variable object

imms_var_list_destroy()   destroys an IMMS variable object

imms_var_list_add_var()   add an IMMS variable object to an IMMS variable list

imms_var_list_delete_var()   deletes an IMMS variable object from an IMMS variable list object

imms_var_list_get_nth_var()   returns the nth variable object in a variable list

Creating, accessing, and destroying variable types

imms_type_create()   creates an IMMS data type object

imms_type_destroy()   destroys an IMMS data type object

imms_type_is_deletable()   informs whether a variable type can be delected from a device

imms_type_lookup_dl()   looks up a type's dl-string in the ISCU database

Managing devices

imms_identify()   obtains identifying information about the device

imms_read_capability_list()   gets the list of device capabilities

imms_read_object_names()   gets the list of the objects

imms_read_status()   gets the status of a device

imms_read_dattributes()   gets the attributes of a domain

Managing domains

imms_delete_domain()   deletes a domain from the specifies device

imms_download_domain()   downloads a domain to the specified device from a file on the client's computer

imms_read_attributes()   gets the domain attributes

imms_upload_domain()   uploads a domain from the specified device to a host file

imms_domain_set_sharable()   sets the share status of a domain

Managing the device's environment

imms_abort()   attempts to abruptly terminate an association
                                           between the application and a device

imms_cancel_request()   cancels a confirmed request to a device

imms_conclude()   attempts to gracefully terminate an association
                                           between an application and a device

imms_initiate()   initiates an association with a device

imms_detect_device_term()   gives notification when an association with a
                                           device is terminated

Accessing variables in devices

imms_define_type()   defines a named–variable–type on the specified
                                           device

imms_define_var()   defines a variable on the specified device

imms_define_var_list()   defines a variable list on the specified device

imms_delete_var()   deletes a variable from the specified device

imms_delete_var_list()   deletes a variable list from the specified device

imms_delete_type()   deletes a type from the specified device

imms_read_var_list()   reads a variable list from the specified device

imms_read_vattributes()   gets the attributes of a variable from the device
                                           where the variable is defined

imms_read_tattributes()   gets the attributes of an MMS data type

imms_write_var_list()   writes a variable list to the specified device

Information Reports

imms_disarm_info_reports()   disables the receipt of information reports from a
                                           device

imms_arm_info_reports()   enable the receipt of information reports from a
                                           device

imms_var_disarm_info_reports()

imms_device_get_info_status()   gets the status of the information reports service on
                                           the device

Named Variable Access

imms_name_list_created()   creates an IMMS name list object associated with
                                           an IMMS device object

imms_name_list_destroy()	destroys an IMMS name list object
imms_name_list_get_nth_name()	returns a pointer to the nth object name in the list of object names defined on the device
imms_read_name_list()	retrieves the list containing the names of the objects defined at the device

3.9. DEVICE DRIVER TOOLKIT CLUSTER (IDDK)

The Device Driver Toolkit (IDDK) provides the programmer with a set of functions that support the development of device drivers for non–MMS devices. IDDK also provides a C–language template for device drivers that the programmer can adapt to create a driver for a particular device.

IDDK provides functions that enable you to create an interface between a virtual manufacturing device and device–independent or device–dependent code, thus enabling non–MMS devices to function on a network as if they were MMS devices.

Specialized queued I/O functions support RS–232 communications.

3.10. DEVICE DRIVERS

MainStream provides device drivers for a number of programmable devices. Each driver acts as a gateway between MMS and the device's proprietary data highway. The driver emulates the MMS Virtual Manufacturing Device (VMD). Each driver works with and requires the MainStream Device Communications option.

3.11. COMPARE LIBRARY FOR PROGRAMMABLE CONTROLLERS (ICLP)

The MainStream Compare Library for Programmable Controllers (ICLP) is a group of compare programs used to verify the contents of programs uploaded from remote stations. In particular, ICLP:

- Verifies the contents of a programmable controller through comparison to other versions of the program or to other programs.
- Supports Allen–Bradley PLC–2, Allen–Bradley PLC–5, and GE Fanuc programmable controllers
- Is available through the MainStream CCA Program and Recipe Management package.

3.12. GLOBAL EVENT LOGGING UTILITY (IELU)

The Global Event Logging Utility (IELU) provides a way to log the occurrence of significant runtime events. IELU enables a programmer to define any type of occurrence – for example, an equipment failure – as an event that is to be logged.

Messages giving notice that an event has occurred can be sent to any number of destinations known as subscribers. Typically, subscribers are devices such as printers, files, or workstations. Each subscriber can receive notice of any number of distinct events from any number of sources.

An application program uses a function from IELU's callable interface to send notice of a detected event, by means of a transaction message, to the Logger Process. The Logger Process then optionally formats and distributed event messages, sending the event to the event's subscribers.

IELU consists of the following components:

- A Configuration Data File, an ASCII file in which you define events, subscribers, and event–subscriber links
- Functions that your MainStream application program uses to communicate with the Logger Process
- A Logger Process that responds to a notice of an event by notifying all subscribers to that event (as defined in the links section of the Configuration Data File)

Log Event Function

ielu_put()   logs an event to the Logger Process

Shut Down Logger Function

ielu_shutdown_request()   shuts down the Logger Process

Event Subscriber Function

ielu_reset_subscriber()   resets a subscriber's log

3.13. GLOBAL ALARM MANAGEMENT UTILITY (IAMU)

The Global Alarm Management Utility (IAMU) enables a process to notify other processes about an alarm event. An alarm is an object possessing a set of attributes that you manipulate to manage the alarm. You define the alarm and some of its attributes.

IAMU maintains a table of alarms. The Event Logger Utility can distribute notices of these alarms to a set of processes that your application specifies.

IAMU provides a library of callable functions that enable an application program to:

- Set an alarm
- Turn off, or clear, an alarm
- Acknowledge an alarm
- Set the priority of an alarm

IAMU also provides an Alarm Monitor Utility program that enables a user to view, acknowledge, and clear alarms. The user can choose to view only those alarms that have to do with a given sphere of interest or responsibility.

Active Alarm Functions

iamu_alarm_request_ack()	acknowledges an active alarm
iamu_alarm_request_clr()	clears an active alarm
iamu_alarm_request_modpri()	changes an active alarm's priority attribute

iamu_alarm_request_set()sets an alarm

Alarm Manager Functions

iamu_shutdown_request()	shuts down the Alarm Manager process

3.14. TEST SIMULATOR UTILITY (ITSU)

The MainStream Test Simulator Utility (ITSU) enables you to test one or more processes in an industrial automation system. ITSU consists of standard functions that you can use to build an interactive Test Simulator Utility, and other software tools that enable you to create a simulated test environment.

In particular, the Test Simulator Utility enables you to:

- Test a single part of a complex system before the entire system is programmed
- Simulate conditions that will be created by processes that are to interface with the process that you want to test.

- Observe each response of the process being tested in real time
- Change the dynamics of the test at any time during the test
- Record each response of the process being tested for further study
- Create a test clock according to which time passes more quickly or more slowly than actual time. A process under test can be made to run according to test–clock time. ITSU uses services of the Time Management Cluster (ITIM) to create a test clock.

3.15. SYSTEM CONFIGURATION UTILITY (ISCU)

The MainStream System Configuration Utility (ISCU) defines, or configures, data that helps regulate how your MainStream application program accesses system resources. Configuring resources guarantees that:

- Conflicts do not arise among processes that require the same types of resources
- Enough of these resources are reserved for the use of your MainStream application program

Configuration data can refer both to resources of the operating system and to resources provided by MainStream itself. It can refer both to local resources, such as tables and data structures, and to physical resources, such as shared memory. ISCU stores configuration data in an ISCU database. The ISCU database is stored in shared memory for runtime access.

INTEGRA PROJECT MEMBERS UNDER EUREKA:

AVENIR / Noratom A/S, Oslo, Norway
DANFOSS A/S, Nordborg, Denmark
IKOSS GmbH, Stuttgart, Germany
ITP Automazione SpA, Torino, Italy
Nordic Manufacturing Forum / Nordic Enterprise A/S, Oslo, Norway
SEMA METRA SA, Paris, France
VALMET Oy, Helsinki, Finland

Members of the Nordic Manufacturing Forum:

Alfa Laval, Sweden
Avenir / Noratom, Norway
Danfoss, Denmark
DTH, Denmark (assoc. member)
Kverneland, Norway
Jordan, Norway
Nokia, Finland
Nordic Enterprise, Norway (Secretariat)
Sandvik Coromant, Sweden
S.T. Lyngsø, Denmark
Valmet, Finland
Volvo, Sweden

AUTHOR INDEX